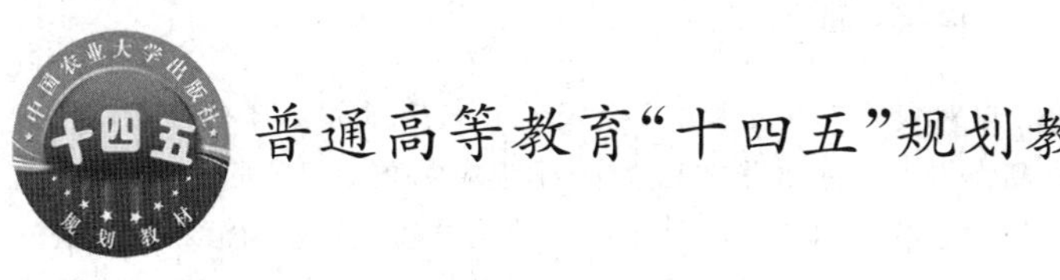

食品学科本科专业英语

第3版

陈宗道　刘　雄　主编

史贤明　主审

中国农业大学出版社

·北京·

内 容 简 介

具备一定的食品专业外语听说读写能力，是对食品企业技术和安全管理工作者、食品科学研究工作者和食品类专业师生的一项基本要求。因而，食品类专业人才培养方案规定，学生应掌握一门外国语，具有国际视野和跨文化交流、竞争与合作的能力。本教材系统介绍了食品科技论文的英语基本专业词汇和常见语法现象，希望通过大量阅读和练习，提高学生的英语阅读和撰写食品科技论文的能力。本教材内容涉及食品化学、食品微生物学、食品营养学、食品加工工程、食品贮藏工程、食品物流学、食品质量和安全管理、食品包装和食品法规与标准等领域。另外，考虑到部分学生的实际需要，增加了出国留学申请、缩略词和听力提高等章节。

本书可作为食品类6个专业（食品科学与工程专业、食品质量与安全专业、粮食工程专业、乳品工程专业、酿酒工程专业及葡萄与葡萄酒工程专业）本科生的学习教材，也可供硕士生、食品企业和食品贸易工作者学习参考。

图书在版编目(CIP)数据

食品学科本科专业英语 / 陈宗道，刘雄主编. —3版. —北京：中国农业大学出版社，2021.8
ISBN 978-7-5655-2567-4

Ⅰ.①食… Ⅱ.①陈…②刘… Ⅲ.①食品科学-英语-高等学校-教材 Ⅳ.①TS201

中国版本图书馆CIP数据核字(2021)第118969号

书　名 食品学科本科专业英语　第3版
作　者 陈宗道　刘　雄　主编　　史贤明　主审

策划编辑	宋俊果　王笃利　魏　巍	**责任编辑**	赵　艳　魏　巍
封面设计	郑　川		
出版发行	中国农业大学出版社		
社　址	北京市海淀区圆明园西路2号	**邮政编码**	100193
电　话	发行部 010-62733489，1190		读者服务部 010-62732336
	编辑部 010-62732617，2618		出　版　部 010-62733440
网　址	http://www.caupress.cn	**E-mail**	cbsszs@cau.edu.cn
经　销	新华书店		
印　刷	涿州市星河印刷有限公司		
版　次	2021年11月第3版　　2021年11月第1次印刷		
规　格	787×1092　16开本　16.5印张　410千字		
定　价	45.00元		

图书如有质量问题本社发行部负责调换

普通高等学校食品类专业系列教材

编审指导委员会委员

（按姓氏拼音排序）

第3版编审人员

主　　编　陈宗道　刘　雄（西南大学）

副 主 编　（以姓氏笔画为序）

王晓闻（山西农业大学）
刘四新（海南大学）
刘承初（上海海洋大学）
纪淑娟（沈阳农业大学）
李巨秀（西北农林科技大学）
陈　霞（内蒙古农业大学）
周爱梅（华南农业大学）
贺稚非（西南大学）
秦　文（四川农业大学）
聂乾忠（湖南农业大学）
崔　艳（天津农学院）
景　浩（中国农业大学）

编写人员　（以姓氏笔画为序）

王晓闻（山西农业大学）
王淑培（南宁师范大学）
冯　颖（沈阳农业大学）
刘　雄（西南大学）
刘四新（海南大学）
刘承初（上海海洋大学）
纪淑娟（沈阳农业大学）
杜木英（西南大学）
李巨秀（西北农林科技大学）
李从发（海南大学）
陈　霞（内蒙古农业大学）
陈运娇（华南农业大学）
陈宗道（西南大学）
周爱梅（华南农业大学）
贺稚非（西南大学）
秦　文（四川农业大学）
聂乾忠（湖南农业大学）
郭庆启（东北林业大学）
郭　瑜（山西农业大学）
崔　艳（天津农学院）
景　浩（中国农业大学）

主　　审　史贤明（上海交通大学）

第2版编审人员

主　　编　陈宗道　刘　雄（西南大学）

副 主 编　（以姓氏笔画为序）

王晓闻（山西农业大学）
刘四新（海南大学）
刘承初（上海海洋大学）
纪淑娟（沈阳农业大学）
李巨秀（西北农林科技大学）
陈　霞（内蒙古农业大学）
周爱梅（华南农业大学）
贺稚非（西南大学）
秦　文（四川农业大学）
聂乾忠（湖南农业大学）
崔　艳（天津农学院）
景　浩（中国农业大学）

编写人员　（以姓氏笔画为序）

王晓闻（山西农业大学）
王淑培（武夷学院）
冯　颖（沈阳农业大学）
刘　雄（西南大学）
刘四新（海南大学）
刘承初（上海海洋大学）
纪淑娟（沈阳农业大学）
李巨秀（西北农林科技大学）
李从发（海南大学）
陈　霞（内蒙古农业大学）
陈运娇（华南农业大学）
陈宗道（西南大学）
周爱梅（华南农业大学）
贺稚非（西南大学）
秦　文（四川农业大学）
聂乾忠（湖南农业大学）
郭　瑜（山西农业大学）
郭庆启（东北林业大学）
崔　艳（天津农学院）
景　浩（中国农业大学）

主　　审　史贤明（上海交通大学）

出版说明
（代总序）

岁月如梭，食品科学与工程类专业系列教材自启动建设工作至现在的第4版或第5版出版发行，已经近20年了。160余万册的发行量，表明了这套教材是受到广泛欢迎的，质量是过硬的，是与我国食品专业类高等教育相适宜的，可以说这套教材是在全国食品类专业高等教育中使用最广泛的系列教材。

这套教材成为经典，作为总策划，我感触颇多，翻阅这套教材的每一科目、每一章节，浮现眼前的是众多著作者们汇集一堂倾心交流、悉心研讨、伏案编写的景象。正是大家的高度共识和对食品科学类专业高等教育的高度责任感，铸就了系列教材今天的成就。借再一次撰写出版说明（代总序）的机会，站在新的视角，我又一次对系列教材的编写过程、编写理念以及教材特点做梳理和总结，希望有助于广大读者对教材有更深入的了解，有助于全体编者共勉，在今后的修订中进一步提高。

一、优秀教材的形成除著作者广泛的参与、充分的研讨、高度的共识外，更需要思想的碰撞、智慧的凝聚以及科研与教学的厚积薄发。

20年前，全国40余所大专院校、科研院所，300多位一线专家教授，覆盖生物、工程、医学、农学等领域，齐心协力组建出一支代表国内食品科学最高水平的教材编写队伍。著作者们呕心沥血，在教材中倾注平生所学，那字里行间，既有学术思想的精粹凝结，也不乏治学精神的光华闪现，诚所谓学问人生，经年积成，食品世界，大家风范。这精心的创作，与敷衍的粘贴，其间距离，何止云泥！

二、优秀教材以学生为中心，擅于与学生互动，注重对学生能力的培养，绝不自说自话，更不任凭主观想象。

注重以学生为中心，就是彻底摒弃传统填鸭式的教学方法。著作者们谨记“授人以鱼不如授人以渔”，在传授食品科学知识的同时，更启发食品科学人才获取知识和创造知识的思维与灵感，于润物细无声中，尽显思想驰骋，彰耀科学精神。在写作风格上，也注重学生的参与性和互动性，接地气，说实话，“有里有面”，深入浅出，有料有趣。

三、优秀教材与时俱进，既推陈出新，又勇于创新，绝不墨守成规，也不亦步亦趋，更不原地不动。

首版再版以至四版五版，均是在充分收集和尊重一线任课教师和学生意见的基础上，对新增教材进行科学论证和整体规划。每一次工作量都不小，几乎覆盖食品学科专业的所有骨干课程和主要选修课程，但每一次修订都不敢有丝毫懈怠，内容的新颖性，教学的有效性，齐头并进，一样都不能少。具体而言，此次修订，不仅增添了食品科学与工程最新发展，又以相当篇幅强调食品工艺的具体实践。每本教材，既相对独立又相互衔接互为补充，构建起系统、完整、实用的课程体系，为食品科学与工程类专业教学更好服务。

四、优秀教材是著作者和编辑密切合作的结果，著作者的智慧与辛劳需要编辑专业知识和奉献精神的融入得以再升华。

同为他人作嫁衣裳，教材的著作者和编辑，都一样的忙忙碌碌，飞针走线，编织美好与绚丽。这套教材的编辑们站在出版前沿，以其炉火纯青的编辑技能，辅以最新最好的出版传播方式，保证了这套教材的出版质量和形式上的生动活泼。编辑们的高超水准和辛勤努力，赋予了此套教材蓬勃旺盛的生命力。而这生命力之源就是广大院校师生的认可和欢迎。

第1版食品科学与工程类专业系列教材出版于2002年，涵盖食品学科15个科目，全部入选“面向21世纪课程教材”。

第2版出版于2009年，涵盖食品学科29个科目。

第3版(其中《食品工程原理》为第4版)500多人次80多所院校参加编写，2016年出版。此次增加了《食品生物化学》《食品工厂设计》等品种，涵盖食品学科30多个科目。

需要特别指出的是，这其中，除2002年出版的第1版15部教材全部被审批为“面向21世纪课程教材”外，《食品生物技术导论》《食品营养学》《食品工程原理》《粮油加工学》《食品试验设计与统计分析》等为“十五”或“十一五”国家级规划教材。第2版或第3版教材中，《食品生物技术导论》《食品安全导论》《食品营养学》《食品工程原理》4部为“十二五”普通高等教育本科国家级规划教材，《食品化学》《食品化学综合实验》《食品安全导论》等多个科目为原农业部“十二五”或农业农村部“十三五”规划教材。

本次第4版(或第5版)修订，参与编写的院校和人员有了新的增加，在比较完善的科目基础上与时俱进做了调整，有的教材根据读者对象层次以及不同的特色做了不同版本，舍去了个别不再适合新形势下课程设置的教材品种，对有些教材的题目做了更新，使其与课程设置更加契合。

在此基础上，为了更好满足新形势下教学需求，此次修订对教材的新形态建设提出了更高的要求，出版社教学服务平台“中农 De 学堂”将为食品科学与工程类专业系列教材的新形态建设提供全方位服务和支持。此次修订按照教育部新近印发的《普通高等学校教材管理办法》的有关要求，对教材的政治方向和价值导向以及教材内容的科学性、先进性和适用性等提出了明确且具针对性的编写修订要求，以进一步提高教材质量。同时为贯彻《高等学校课程思政建设指导纲要》文件精神，落实立德树人根本任务，明确提出每一种教材在坚持食品科学学科专业背景的基础上结合本教材内容特点努力强化思政教育功能，将思政教育理念、思政教育元素有机融入教材，在课程思政教育润物细无声的较高层次要求中努力做出各自的探索，为全面高水平课程思政建设积累经验。

教材之于教学，既是教学的基本材料，为教学服务，同时教材对教学又具有巨大的推动作用，发挥着其他材料和方式难以替代的作用。教改成果的物化、教学经验的集成体现、先进教学理念的传播等都是教材得天独厚的优势。教材建设既成就了教材，也推动着教育教学改革和发展。教材建设使命光荣，任重道远。让我们一起努力吧！

罗云波

2021 年 1 月

第 3 版前言

中国农业大学出版社的教材《食品学科本科专业英语》第 1 版出版于 2009 年，2016 年修订增添部分内容后，转眼又过去 5 年时光。在“十三五”期间，我国食品工业保持了良好的增长势头，2019 年我国规模以上企业食品工业营业收入达到 8.1 万亿元，保持了国内第二大支柱产业的地位，有效地保证了人民群众的食品消费需求。面对 2020 年的新冠疫情，食品工业成为经济复苏的强劲动力。食品工业的发展要求教育部门培养一大批高素质的创新人才。食品学科专业英语是学习、了解和把握国内外食品科技发展的重要语言工具，在推动创新人才培养、开展国际交流、实施“一带一路”倡议中具有重要意义。“全球化”是世界各国通过贸易、人员、资金等的密切往来，使各国经济相互依存、促进整体发展、实现凝聚的过程。面对当前“全球化”和“逆全球化”之争愈演愈烈的形势，我们必须认识到全球化是人类历史发展的必然趋势，是社会生产力发展的客观要求和科技进步的必然结果。教育工作者和学生都要对国际形势和未来发展做出正确判断，增强学习专业英语的信心。

2016 年 12 月 8 日，习近平总书记在全国高校思想政治工作会议上强调：“要坚持把立德树人作为中心环节，把思想政治工作贯穿教育教学全过程，实现全程育人、全方位育人，努力开创我国高等教育事业发展新局面。”《食品学科本科专业英语》一书的修订工作正是根据这一指示精神，结合食品产业和食品专业英语教学工作的发展，在中国农业大学出版社编辑们不懈努力下进行的。编者在教材内容的取材上注重博大精深的中华饮食文化、国内食品科技工作者的科研成果等，强化树立“道路自信、理论自信、制度自信、文化自信”的四个自信意识，注重政治认同和家国情怀的培养。编者希望各位教师、学生在使用本教材时把立德树人的政治思想工作贯彻始终。

此次修订工作得到了许多院校和教师的支持，所有编者都很认真负责，力争提高修订版教材的质量，尽力做到让使用教材的老师和同学满意。

本版教材修订了原版内容的不规范之处，对原来部分教材内容译文进行完善，删除调整了部分内容，由原来 15 章变为 13 章。修订工作仍然坚持两个编写原则：一是选材要顾及整个学科范围，尽可能涉及食品化学、食品微生物学、食品营养学、食品加工工程、食品贮藏工程、食品物流学、食品质量和安全管理、食品包装、食品法规与标准和信息管理等领域；二是难度适中，做到专业上不要太高深，语言上不要太生涩。

本版教材共分 13 章：第 1 章到第 9 章为主体部分，教师和学生可从中选择精读和泛读内容；第 10 章到第 13 章为辅助支持部分，帮助学生全面提升专业英语综合能力。

第 3 版教材各章编者情况如下：第 1 章文摘，由湖南农业大学聂乾忠编写；第 2 章综述，由西南大学刘雄、杜木英和中国农业大学景浩编写；第 3 章教材，由沈阳农业大学纪淑娟、冯颖及华南农业大学周爱梅和山西农业大学郭瑜编写；第 4 章论文的题目、摘要和关键词，由海南大学李从发、刘四新编写；第 5 章论文的前言，由西南大学刘雄和东北林业大学郭庆启编写；第 6 章论文的材料与方法，由山西农业大学王晓闻编写；第 7 章论文的结果与讨论，由西北农林科技大学李巨秀编写；第 8 章论文的结论，由天津农学院崔艳和华南农业大学陈运娇编写；第 9 章参考文献，由四川农业大学秦文编写；第 10 章食品科技写作常用句型，由内蒙古农业大学陈霞编写；第 11 章英文食品科技信息的获得，由西南大学陈宗道、刘雄和南宁师范大学王淑培编写；第 12 章出国留学申请，由上海海洋大学刘承初编写；第 13 章缩略词，由西南大学贺稚非编写；附录作业参考答案由以上各位编者共同编写。全书由陈宗道、刘雄统稿。

编者坦承水平有限，书中错误在所难免。如您发现错误，请不吝赐教，待本书再次修订时改正。

编　者

2020 年 9 月 20 日

第 2 版前言

中国农业大学出版社的教材《食品学科本科专业英语》出版于 2009 年，距今已七载，弹指一挥间。6 年中，我国食品工业有很大的发展，2008 年我国食品工业总产值为 7.7 万亿元，到 2014 年已为 12 万亿元。中国食品产业已经成为国内第二大支柱产业，有效地保证了人民群众的食品消费需求。食品工业的发展要求教育部门培养一大批高素质的创新人才。2009 年我国发布了《食品安全法》，2015 年又对《食品安全法》进行了修订，我国食品安全总体形势稳中向好。2015 年英国经济学人智库发布了《全球食品安全指数报告》，中国在 107 个国家中位居 42，列入表现良好一档(good performance)。

当今食品学科本科专业英语教学工作也今非昔比了。主编在 20 世纪 80 年代开始从事专业英语教学工作，到 2015 年仍未完全脱离专业英语的教学工作，发现现在专业英语教学工作的地位、作用、内容和形式都有了很大变化。

首先，大环境变了，英语在政治、经济、科学、教育和社会活动中的作用越来越重要了，英语能力在学生能力和素质评价体系中的权重提高了。第二，专业英语在食品本科教育中占据了一定的位置。全国有近 300 所高等院校开设了食品类专业，其中绝大多数院校都开设了食品学科本科专业英语课程。食品类专业有 6 个专业：食品科学与工程专业 0827-01(268 所院校开设)、食品质量与安全专业 0827-02(178 所院校开设)、粮食工程专业 0827-03、乳品工程专业 0827-04、酿酒工程专业 0827-05 及葡萄与葡萄酒工程专业 0827-06T，都开设了食品学科本科专业英语课程。第三，从事专业英语教学工作的老师多是博硕士毕业，既有较高的基础英语水平，又有较高的食品专业学术水平。学生的基础英语水平和读说听写能力也比以往高了许多，并有学习专业英语的内在要求和强烈愿望，能积极主动地学习。第四，过去专业英语资料仅限于图书馆屈指可数的纸质英语教材、专著和杂志，而现在网上的专业英语资料浩如烟海。过去听和说专业英语的机会甚少，仅限于外国专家来校做学术报告，而现在相当一部分院校使用英语原版书进行教学，课堂上用全英文授课或双语教学，网上还有慕课、优酷等网站有英语教学视频及英语配音的食品加工和食品物流等的视频。第五，早期专业英语教学没有专用的专业英语教材，而现在各出版社都有可供选择的专业英语教材。第六，专业英语教学在本科教育中发挥了更大的作用，提高了学校国际化水平，增强了学生获取和使用英语学术资源的能力及与国际同行进行学术交流的能力，提高了本科毕业论文的质量水平。

食品产业和食品专业英语教学工作的发展，要求修订《食品学科本科专业英语》教材。这次修订工作得到了许多院校和老师的支持，纷纷要求加盟编写队伍。修订工作又吸纳了 6 位老师加入，为修订版教材增添了光彩。所有编者都很认真负责，力争提高修订版教材的质量，尽力做到让使用教材的老师和同学双满意。

修订版教材保持原版的框架结构不变，保持字数基本不变，删除了部分老旧的内容，增添了新的内容。增添了新的两章：第 11 章专业英语常用句型，第 12 章提高专业英语听力的途径和资源。

修订工作仍然坚持两个编写原则：一是选材要顾及整个专业面，尽可能涉及食品化学、食品微生物学、食品营养学、食品加工工程、食品贮藏工程、食品物流学、食品质量和安全管理、食品包装、食品法规与标准和信息管理等领域。二是掌握难度适中，做到专业上不要太高深，语言上不要太生涩。应出版社的统一要求，修订版教材增加了二维码，引导学生应用网络资源进行学习，帮助学生扩展学习面。

修订版教材共分 15 章。全教材可分为两部分：第一部分为主体部分，由第 1 章到第 10 章，教师和学生可从中选择精读和泛读内容；第二部分为辅助支持部分，由第 11 章到第 15 章，帮助学生全面提升读说听写能力，如缩略词有助于读，食品科技写作常用句型和出国留学申请有助于写，提高专业英语听力的途径和资源有助于听说。

修订版教材的各章及其编者如下：第 1 章文摘，由湖南农业大学聂乾忠编写。第 2 章综述，由西南大学刘雄和中国农业大学景浩编写。第 3 章教材，由沈阳农业大学纪淑娟、冯颖及华南农业大学周爱梅和山西农业大学郭瑜编写。第 4 章论文的题目、摘要和关键词，由海南大学李从发、刘四新编写。第 5 章论文的前言，由西南大学刘雄编写。第 6 章论文的材料与方法，由山西农业大学王晓闻编写。第 7 章论文的结果与讨论，由西北农林科技大学李巨秀编写。第 8 章论文的结论，由天津农学院崔艳和华南农业大学陈运娇编写。第 9 章参考文献，由四川农业大学秦文编写。第 10 章新闻报道，由西南大学陈宗道和东北林业大学郭庆启编写。第 11 章食品科技写作常用句型，由内蒙古农业大学陈霞编写。第 12 章提高专业英语听力的途径和资源，由武夷学院王淑培编写。第 13 章英文食品科技信息的获得，由西南大学陈宗道、刘雄编写。第 14 章出国留学申请，由上海海洋大学刘承初编写。第 15 章缩略词，由西南大学贺稚非编写。

编者坦承水平极其有限，书中错误在所难免。如您发现错误，请不吝赐教，待本书再次修订时改正。

编　者

2015 年 11 月 20 日

主审感言

汉语和英语是世界上应用最广泛的两种语言，它们分别又是东、西方具有代表性的语言，因此，这两种语言的互译以及学习它们之间互译的方法十分重要。同时，中、英两种语言都具有悠久的历史和深厚的文化背景。因此，准确地完成这两种语言互译或转换并非是一件容易的事情。

食品产业已逐渐成为一种国际化的产业，食品国际贸易的迅速发展使世界各国人民能共享不同民族的劳动成果。同时，也使世界各国人民共同关注食品安全问题。因此，食品经济和饮食文化的这种国际化发展趋势对语言的交流提出了更高的要求。

近年来，我国的食品科学与工程及食品质量与安全两专业得到了飞速发展，这两个专业涉及的新概念和新名词及其外来语不断涌现。因此，食品专业的学生对专业英语的学习十分渴望。然而这两个专业的专业英语教师和教材都较匮乏，陈宗道教授和刘雄教授主编的这本教材在这种背景下孕育而生，十分及时和必要。这本教材具有较强的系统性，覆盖了两个专业的主要内容，包括了中英文互译中的常见问题。此外，范文的翻译较规范。这些对学生学习专业英语是颇有帮助的。

由于东西方民族在思维方式和文化上存在差异，由于不同的人在理解和文字表达上存在差异，所以，一些学术术语的中英文互译会有差异，这种差异的存在和发展会阻碍食品科技的国际交流，甚至影响学术的发展。因此，同行之间加强沟通，尽量达成共识，谨慎出炉新名称是科学家、教师和学生在专业英语学习和应用中应该予以倡导的。

目　录

第 1 章　文摘　Abstract ………………………………………………… 1

第 1 课　The pediatric burden of rotavirus disease in Europe ………………… 2
轮状病毒引起的幼儿疾病给欧洲造成的负担

第 2 课　Safety and keeping quality of pasteurized milk under refrigeration ………… 3
冷藏条件下巴氏消毒奶的安全性和保质期

第 3 课　Public health significance of antimicrobial-resistant Gram-negative bacteria in raw bulk tank milk ………………………………………… 4
储奶罐生乳中具有耐药性的革兰氏阴性菌的公共健康意义

第 4 课　Non-enzymatic degradation of citrus pectin and pectate during prolonged heating: effects of pH, temperature, and degree of methyl esterification … 5
在延长加热过程中 pH、温度和甲酯化程度对柑橘果胶和果胶物质非酶降解的影响

第 5 课　Effects of fish hem protein structure and lipid substrate composition on hemoglobin-mediated lipid oxidation ………………………………… 7
鱼肉血红素蛋白结构和脂肪组成对血红蛋白调节的脂肪氧化的影响

第 6 课　Comparative whole-grain intake of British adults in 1986—1987 and 2000—2001 ………………………………………………………… 8
1986—1987 年度和 2000—2001 年度英国成年人全谷物摄入量比较

第 7 课　Effect of high-pressure treatment on survival of *Escherichia coli* O157:H7 population in tomato juice ………………………………………… 10
高压处理对番茄汁中大肠杆菌 O157:H7 存活的影响

第 8 课　Improvement of the overall quality of the table grapes stored under modified atmosphere packaging in combination with natural antimicrobial compounds ………………………………………………………… 11
气调包装结合天然抗菌剂能有效改善鲜食葡萄在贮藏中的整体品质

第 9 课　Optimization of an enzyme assisted process for juice extraction and clarification from litchis(*Litchi chinensis* Sonn.) ………………………… 13

在荔枝(*Litchi chinensis* Sonn.)汁的榨取和澄清中酶法反应条件的优化
作　业　汉译英 …… 14
第2章　综述　Review …… 16
第1课　Hazards of packaging materials in contact with foods …… 17
与食品接触的包装材料的危害性
第2课　Iron deficiency and iron fortified foods …… 19
铁缺乏及铁强化食品
第3课　Enzymatic deacidification/reesterification …… 21
酶法脱酸和再酯化
第4课　Physico-chemical properties of modified starch …… 23
变性淀粉的理化性质
第5课　Preservation of wines …… 24
葡萄酒的保存
第6课　Effect of high pressure (HP) on microorganisms in milk …… 26
高压处理对牛奶中微生物的影响
第7课　Subcritical water extraction …… 29
亚临界水萃取
第8课　Maillard reaction …… 32
美拉德反应
作　业　汉译英 …… 35
第3章　教材　Teaching Material …… 36
第1课　General processing concepts …… 37
基本加工概念
第2课　Membrane separations …… 39
膜分离
第3课　Food-parking materials and forms …… 41
食品包装材料和形式
第4课　Description of commercial sterilization systems …… 44
商业杀菌系统的类型
第5课　Food freezing and product quality …… 47
食品冷冻和产品质量
第6课　Introduction to food additives …… 49
食品添加剂概述
第7课　Major chemical processes of food deterioration …… 53
食品腐败的主要化学过程

第 8 课　The nutrients in foods …… 55
食物中的营养素
第 9 课　Sensory evaluation concepts …… 58
感官评价概念
作　业　汉译英 …… 61
第 4 章　论文的题目、摘要和关键词　Title, Abstract and Keywords of Article …… 63
第 1 课　A new method of HACCP for the catering and food service industry …… 64
一种应用于餐饮业的 HACCP 新方法
第 2 课　Incidence and characterization of *Salmonella* species in street food and clinical samples …… 65
沙门菌在街头食品和临床样品中的发生率及特征
第 3 课　The antimicrobial effect of thyme essential oil, nisin, and their combination against *Listeria monocytogenes* in minced beef during refrigerated storage …… 67
百里香精油、乳酸链球菌素及其混合物对冷藏牛肉糜中单核细胞增生李斯特菌的抗菌效果
第 4 课　High hydrostatic pressure processing of fruit and vegetable products …… 68
果蔬产品的高静水压加工
第 5 课　Osmotic dehydration of pineapple as a pre-treatment for further drying …… 70
菠萝渗透脱水作为进一步干燥的预处理
第 6 课　Free-radical-scavenging activity and total phenols of noni (*Morinda citrifolia* L.) juice and powder in processing and storage …… 71
诺丽(*Morinda citrifolia* L.)果汁和果粉的自由基清除能力和总酚含量在加工和贮藏中的变化
第 7 课　Synthesis and characterization of canola oil-stearic acid-based trans-free structured lipids for possible margarine application …… 72
可能应用于人造奶油的菜籽油硬脂酸基无反式脂肪酸的结构油脂的合成及其特性
第 8 课　Fractionating soybean storage proteins using Ca^{2+} and $NaHSO_3$ …… 74
用 Ca^{2+} 和 $NaHSO_3$ 分步分离大豆贮藏蛋白
第 9 课　Quality changes in fresh-cut peach and nectarine slices as affected by cultivar, storage atmosphere and chemical treatments …… 76
品种、贮藏气体和化学处理对鲜切桃片和油桃片品质的影响
第 10 课　Pectinolytic enzymes secreted by yeasts from tropical fruits …… 77
热带水果酵母分泌的果胶酶
作　业　汉译英 …… 78

第 5 章　论文的前言　**Introduction of Article** …… 80

第 1 课　Reduced and high molecular weight barley beta-glucans decrease plasma total and non-HDL-cholesterol in hypercholesterolemic Syrian golden hamsters …… 81
低分子量和高分子量的大麦 β-葡聚糖降低患高胆固醇血症的叙利亚金鼠血中总胆固醇和非高密度脂蛋白胆固醇浓度

第 2 课　Optimization of the jet steam instantizing process of commercial maltodextrins powders …… 83
商品糊精粉的喷流预处理工艺的优化

第 3 课　Sterilization solutions for aseptic processing using a continuous flow microwave system …… 86
用于连续微波处理系统消毒的杀菌液

第 4 课　Effect of fermentation temperature and culture media on the yeast lipid composition and wine volatile compounds …… 88
发酵温度和培养基对酵母脂肪组成和葡萄酒香气组成的影响

第 5 课　Development and assessment of pilot food safety educational materials and training strategies for Hispanic workers in the mushroom industry using the Health Action Model …… 92
开发与评估食品安全教育的材料,以及对蘑菇厂的西班牙工人进行健康行为规范的培训策略

第 6 课　Determination of O_2 and CO_2 transmission rates through microperforated films for modified atmosphere packaging of fresh fruits and vegetables …… 94
用于新鲜蔬菜和水果气调包装的微孔膜的 O_2 和 CO_2 透过率的测定

作　业　汉译英 …… 97

第 6 章　论文的材料与方法　**Materials and Methods of Article** …… 99

第 1 课　Hypobaric storage removes scald-related volatiles during the low temperature induction of superficial scald of apples …… 100
减压贮藏可以去除低温导致苹果表皮烫伤过程中烫伤相关的挥发物

第 2 课　Determination of phenolic compounds and their antioxidant activity in fruits and cereals …… 101
水果和谷物中酚类物质的测定及其抗氧化活性研究

第 3 课　Real-time multiplex SYBR Green Ⅰ-based PCR assay for simultaneous detection of *Salmonella* serovars and *Listeia moncytogenes* …… 104
用于同步检测 *Salmonella* serovars 和 *Listeia moncytogenes* 的实时多元 SYBR Green Ⅰ型 PCR

第 4 课 Changes in aroma characteristics of simulated beef flavour by soy protein isolate assessed by descriptive sensory analysis and gas chromatography …… 107
用感官分析和气相色谱评价大豆分离蛋白仿生牛肉的风味品质的变化
第 5 课 Effect of processing on buckwheat phenolics and antioxidant activity …… 109
加工对荞麦酚及其抗氧化活性的影响
第 6 课 Changes in red wine soluble polysaccharide composition induced by malolactic fermentation …… 111
苹果酸发酵引起的红葡萄酒中可溶性多糖的变化
作 业 汉译英…… 112

第 7 章 论文的结果与讨论 Results and Discussion of Article …… 114

第 1 课 The effect of steeping time on the final malt quality of buckwheat …… 115
浸麦时间对荞麦麦芽品质的影响
第 2 课 Volatile and non-volatile chemical composition of the white guava fruit (*Psidium guajava*) at different stages of maturity …… 118
不同成熟阶段的白番石榴中的挥发性和非挥发性化学成分
第 3 课 Sugars, organic acids, phenolic composition and antioxidant activity of sweet cherry (*Prunus avium* L.) …… 121
甜樱桃(*Prunus avium* L.)中糖、有机酸、酚类物质和抗氧化活性
第 4 课 Evaluation of processing qualities of tomato juice induced by thermal and pressure processing …… 123
热加工和高压加工对番茄汁加工品质的影响
第 5 课 Changes of pesticide residues in apples during cold storage …… 126
苹果在冷藏过程中农药残留的变化
作 业 汉译英…… 128

第 8 章 论文的结论 Conclusion of Article …… 130

第 1 课 Rheological behaviour of dairy products as affected by soluble whey protein isolate …… 131
可溶性的乳清蛋白对乳制品的流变学性质的影响
第 2 课 Effect of ethylene in the storage environment on quality of 'Bartlett pears' …… 133
贮藏环境中乙烯对巴特利特梨质量的影响
第 3 课 Effect of nisin on yogurt starter, and on growth and survival of *Listeria monocytogenes* during fermentation and storage of yogurt …… 134
在酸乳发酵和贮藏期中乳酸链球菌素对酸奶发酵剂和单核细胞增生李斯特菌生长的影响

第 4 课 Effect of thermal blanching and of high pressure treatments on sweet green and red bell pepper fruits（*Capsicum annuum* L.）······ 136
热烫和高压处理对青椒和红椒（*Capsicum annuum* L.）的影响

第 5 课 Gelatine-starch films: physicochemical properties and their application in extending the post-harvest shelf life of avocado（*Persea americana*）······ 138
明胶-淀粉膜的理化性质及其在延长鳄梨采后货架期方面的应用

第 6 课 Influence of cooking and microwave heating on microstructure and mechanical properties of transgenic potatoes ······ 139
烹调加热和微波加热对转基因马铃薯微结构和机械特性的影响

第 7 课 Effects of drying process on antioxidant activity of purple carrots ······ 141
干燥过程对紫胡萝卜抗氧化活性的影响

第 8 课 Genetically improved starter strains: opportunities for the dairy industry ······ 142
基因改造的酵母菌株：乳制品工业的机遇

第 9 课 Genetically engineered foods and the environment: a catastrophe in the making ······ 144
基因工程食品与环境：正在形成的灾难

第 10 课 Evolution and stability of anthocyanin-derived pigments during port wine aging ······ 146
波特葡萄酒陈酿期间花青素类色素的稳定性和评价

第 11 课 Acrylamide in cereal products ······ 148
谷类产品中的丙烯酰胺

第 12 课 Nanoemulsion delivery systems for oil-soluble vitamins: influence of carrier oil type on lipid digestion and vitamin D_3 bioaccessibility ······ 150
脂溶性维生素的纳米乳液传递系统：载体油脂类型对油脂消化和维生素 D_3 生物有效性的影响

作 业 汉译英 ······ 152

第 9 章 参考文献 References ······ 154

第 1 课 概述 ······ 155

第 2 课 实例 ······ 163

作 业 汉译英 ······ 168

第 10 章 食品科技写作常用句型 Common Patterns of Food Scientific Thesis ······ 170

第 1 课 摘要（Abstract）······ 171

第 2 课 引言（Introduction）······ 171

第 3 课 材料与方法（Materials and Methods）······ 172

第 4 课 结果与讨论（Results and Discussion）······ 172

第 5 课　结论(Conclusion) …… 174
第 6 课　综述(Review) …… 174
第 7 课　致谢(Acknowledgement) …… 175
第 11 章　英文食品科技信息的获得 Acquirement of English Information of Food Science and Technology …… 176
第 1 课　主要的英文杂志 …… 177
第 2 课　主要的英文数据库 …… 179
第 3 课　专业英语学习的一些资源 …… 180
第 12 章　出国留学申请 Submit Application for Advanced Study Abroad …… 181
第 1 课　概述 …… 182
第 2 课　个人陈述 …… 184
第 3 课　推荐信 …… 192
第 13 章　缩略词　Abbreviation …… 196
第 1 课　食品加工与贮藏类 …… 197
第 2 课　食品微生物学及生物工程类 …… 200
第 3 课　食品化学类 …… 205
第 4 课　食品包装类 …… 215
第 5 课　与食品有关的组织和标准类 …… 217
附录　作业参考答案 …… 224

第 1 章

文　摘
Abstract

第1课　The pediatric burden of rotavirus disease in Europe

1.原文

引自：Food Science and Technology Abstracts,2007,39(1):63

源自：Epidemiology and Infection.2006,134(5):908-916

Rotaviruses are a major cause of hospitalizations for acute gastroenteritis in developed countries. In this article, the burden of rotavirus disease in < 5-year-old children in Europe was investigated[1] based on currently available data. Results estimated that the annual number of hospitalization for community-acquired rotavirus disease in the 23 million under 5 s living in the EU-25 regions was between 72 000 and 77 000, with a median cost of €1 417 per case. Annual hospitalizations incidence rates ranged from 0.3 to 11.9/1 000 children<5 years old (median 3/1 000). The median proportion of hospital-acquired rotavirus disease among all cases of hospitalization for rotavirus disease was estimated to be 21%. Countries in the EU-25 require information regarding the prevalence of rotavirus disease to support introduction of rotavirus vaccines. Data on cases[2] treated at home, medical visits and emergency wards as well as rotavirus-associated deaths were limited. It is concluded that to fully evaluate the impact and effectiveness of rotavirus vaccination program in Europe, additional epidemiological studies are required.

2.词或词组

community-acquired　社区所致的

epidemiological study　流行病学调查

pediatric　*adj.*　小儿科的

prevalence　*n.*　流行

rotavirus-associated death　轮状病毒所致的死亡

rotavirus　*n.*　轮状病毒

3.主要语法现象

[1] 过去分词作方式状语

[2] 过去分词作定语

4.原文的参考译文

第1课　轮状病毒引起的幼儿疾病给欧洲造成的负担

在发达国家,轮状病毒是导致病人急性肠炎住院的主要病因。本文根据目前已有的资料就欧洲5岁以下的儿童由轮状病毒患病所造成的负担进行了调查研究。结果显示:在欧洲25

个国家 2 300 万 5 岁以下的儿童中，每年因社区获得性轮状病毒疾病而住院的为 72 000～77 000 人，平均医疗费用为€1 417/人。5 岁以下的儿童年住院率为(0.3～11.9)/1 000 人(平均为 3/1 000 人)。在所有住院病因中，因轮状病毒生病而需住院的平均百分率预计为 21%。欧洲 25 国迫切需要相关信息来说明因为轮状病毒的高致病性而急需引入轮状病毒疫苗的必要性。有关轮状病毒所致的家庭治疗、医院治疗、急诊以及所致的死亡情况的资料十分有限。由此可以看出，要对欧洲轮状病毒疫苗计划的有效性和相关影响进行充分的评估，还需开展相关的流行病学调查。

第2课　Safety and keeping quality of pasteurized milk under refrigeration

1.原文

引自： Food Science and Technology Abstracts，2007，39(9)：253

源自： Journal of Food Science and Technology. 2007，44(4)：363-366

Assessment was made of the keeping quality of pasteurized milk under refrigeration (4±1) ℃ by evaluating its microbial，sensory and physicochemical properties on days 0，2，4，6，8，10 and 12 of storage. The initial growth rate of microorganisms was slow. Total viable counts and coliform counts of pasteurized milk increased significantly ($P \leqslant 0.01$) from the 4th day onwards，whereas significant increases in faecal streptococcal counts occurred from the 6th day onwards. Increases in *Escherichia coli* counts，psychrotrophic counts and fungal counts between day zero and day 12 were 0.2，3.7 and 2.5 $\log_{10}$ CFU/mL respectively. *E. coli* was isolated from 1 batch of milk on all days of storage；the isolates belonged to serotype O148 and were positive in the Congo red binding test. *Staphylococcus aureus* was isolated from 1 sample (5%) on the 6th day and from 2 sample (10%) on the 8th day of storage. None of samples revealed the presence of *L. monocytogenes*. Sensory analysis revealed that scores for color，appearance，flavor，aroma and body decreased during storage，but milk was graded as fair on the 8th day. Development of salty or stale flavor，off-odor and presence of clotted particles indicated spoilage. There was a reduction in mean pH value throughout the storage period. The samples were positive for clot formation in the boiling test on day 12 of storage but became unacceptable before this point was reached.

2.词或词组

coliform counts　大肠菌群数

Escherichia coli　大肠杆菌

faecal streptococcal count　粪链球菌数

L. monocytogenes　单核细胞增生李斯特杆菌

physicochemical property　物理化学特性

psychrotrophic *adj*. 嗜冷性的

Staphylococcus aureus 金黄色葡萄球菌

3.原文的参考译文

第2课 冷藏条件下巴氏消毒奶的安全性和保质期

对冷藏(4±1) ℃条件下储存0、2、4、6、8、10和12 d后的巴氏消毒奶的微生物学特性、感官特性及物理化学特性进行评价,从而评价巴氏消毒奶的储藏品质。微生物在开始阶段的生长繁殖十分缓慢。储藏4 d后巴氏消毒奶中细菌总数和大肠菌群显著增加($P \leqslant 0.01$),而6 d后粪链球菌的活菌数会显著增加。大肠杆菌,嗜冷菌和真菌数(CFU/mL)0~12 d的增加数分别为0.2、3.7和2.5对数级增长。在同一批次的牛奶中整个储藏期均分离到大肠杆菌,其属于O148血清型,而且刚果红试验为阳性。在储藏6 d的一个样品中(5%)和8 d的2个样品中(10%)分离到金黄色葡萄球菌。在所有样品中均未检出单核细胞增生李斯特杆菌。感官分析的结果表明产品的色泽、外观、风味和香气均随储藏期的延长而下降,但直到第8天仍可以接受。产品中出现咸味或变味、臭味和凝结成块,则表明变质。储藏期中平均pH有所下降。储藏12 d的样品煮沸凝结实验为阳性,但实际上在此之前产品已变得不可接受(不能食用)了。

第3课 Public health significance of antimicrobial-resistant Gram-negative bacteria in raw bulk tank milk

1.原文

引自: Food Science and Technology Abstracts,2007,39(1):250

源自: Foodborne Pathogens and Disease. 2006,3(3):222-223

The dairy farm environment and animals in the farm serve as important reservoirs of pathogenic and commensal bacteria that could potentially gain access to milk in the bulk tank via several pathways. Pathogenic Gram-negative bacteria can gain access to bulk tank milk from infected mammary glands, contaminated udder and milking machines, and /or from the dairy farm environment. Raw milk consumption can result in exposure to antimicrobial-resistant commensal Gram-negative bacteria. This paper examines the prevalence and role of commensal Gram-negative enteric bacteria in bulk tank milk and their public health significance. Aspects considered include: antimicrobial selection pressure; antimicrobial-resistant Gram-negative pathogens in bulk tank milk; antimicrobial-resistant Gram-negative commensal bacteria in bulk tank milk; and transfer of antimicrobial resistance.

2.词或词组

antimicrobial-resistant 耐药性

commensal bacteria　共生菌

Gram-negative bacteria　革兰氏阴性菌

mammary gland　乳腺

3.原文的参考译文

第 3 课　储奶罐生乳中具有耐药性的革兰氏阴性菌的公共健康意义

奶牛场的环境及奶牛是许多致病菌和共生菌的重要来源，它们可以通过各种途径进入储奶罐的牛乳中。这些具有致病性的革兰氏阴性菌可以通过受感染的乳腺，受污染的乳房及挤奶机和(或)奶牛场的环境进入牛乳中。直接饮用未经处理的生乳将使消费者直接暴露在这些共生的具有耐药性的革兰氏阴性菌的作用下。本文研究了储奶罐原料乳中共生的革兰氏阴性菌的作用及公共健康意义。考虑的方面包括：抗生素的选择压力；储奶罐原料乳中耐药性的革兰氏阴性致病菌；储奶罐原料乳中耐药的革兰氏阴性共生菌；耐药性的转移。

第 4 课　Non-enzymatic degradation of citrus pectin and pectate during prolonged heating: effects of pH, temperature, and degree of methyl esterification

1.原文

引自： Food Science and Technology Abstracts，2007，39(10)：16

源自： Journal of Agriculture and Food Chemistry. 2007，55(13)：5131-5136

The underlying mechanisms governing nonenzymic pectin and pectate degradation during thermal treatment have not yet been fully elucidated. This study determined the extent of non-enzymatic degradation due to β-elimination, acid hydrolysis and demethylation during prolonged heating of citrus pectins and its influence on physicochemical properties. The aim of the study is to improve the quality of food products containing pectins by understanding the mechanics of pectin solubilization and degradation and[1] how they may be minimized. Solutions of citrus pectins,[2] buffered from pH 4.0 to 8.5, were heated at 75, 85, 95 and 110 ℃ for 0-300 min. Evolution of methanol and formation of reducing groups and unsaturated uronides were monitored during heating. Molecular weight and viscosity changes were determined through size exclusion chromatography and capillary viscometry, respectively. Results showed that at pH 4.5, the activation energies of acid hydrolysis, β-elimination and demethylation were 95, 136 and 98 kJ/mol, respectively. This means that at this pH, acid hydrolysis occurs more rapidly than β-elimination. Furthermore, the rate of acid hydrolysis is diminished by higher levels of methyl esterification. Also, citrus pectin (93% esterified) degrades primarily via β-elimination even under acidic conditions. Acid hydrolysis and β-elimination caused significant reduction in relative viscosity and molecular weight.

2.词或词组

capillary viscometry　毛细管黏度法
chromatography　*n*.　色谱层析
demethylation　*n*.　去甲基化
evolution of methanol　甲醇的产生
methyl esterification　甲酯化
non-enzymatic degradation　非酶降解
pectin and pectate　果胶和果胶物质

3.主要语法现象

[1] 宾语从句
[2] 过去分词作定语

4.原文的参考译文

第4课　在延长加热过程中pH、温度和甲酯化程度对柑橘果胶和果胶物质非酶降解的影响

迄今为止对热处理过程中主导果胶和果胶物质降解的深层机制并没有彻底阐明。本研究测定了柑橘果胶延长加热时由β-消除作用、酸水解和去甲基化所导致的非酶降解的程度及对物理化学特性的影响。研究目的是掌握果胶物质的溶解和降解的机制,以及如何减小降解程度,从而改善此类含果胶食品的品质。将柑橘果胶物质溶液配成pH 4.0～8.5的缓冲溶液,在75、85、95和110℃条件下加热0～300 min。加热过程中对甲醇和还原基团及不饱和的糖醛酸进行测定。分别通过分子体积排除色谱法和毛细管黏度法来测定分子质量和黏度的变化。结果显示在pH 4.5时,酸水解、β-消除作用和去甲基化的活化能分别为95、136和98 kJ/mol。这意味着在此pH下,酸水解较β-消除作用更先发生,而且高甲酯化程度降低了酸水解的速率。同样,柑橘果胶(93%酯化度)即使在酸性条件下也主要通过β-消除作用进行降解。酸水解和β-消除作用会使果胶相对黏度和分子质量明显下降。

第5课 Effects of fish hemprotein structure and lipid substrate composition on hemoglobin-mediated lipid oxidation

1.原文

引自： Food Science and Technology Abstracts,2007,39(9):281

源自： Journal of Agriculture and Food Chemistry.2007,55(9):3643-3654

Haemoglobin (Hb) promoted lipid oxidation more effectively in washed tilapia[1] as compared to washed cod in spite of a 2.8-fold higher polyenoic index in the washed cod. This suggested that increasing the fatty acid unsaturation of the substrate did not accelerate the onset of lipid oxidation. Substantial phospholipid hydrolysis in the washed cod was observed,[2] which has the potential to inhibit lipid oxidation. MetHb formation and lipid oxidation occurred more rapidly at pH 6.3 as compared to pH 7.4. Trout Hb autoxidized faster and was a better promoter of lipid oxidation as compared to tilapia Hb. The greater ability of trout Hb to promote lipid oxidation was attributed in part to its lower conformational and structural stability[3] based on secondary and tertiary structure, acid-inducing unfolding, and thermal aggregation measurements. It is suggested that the structural instability and lipid oxidation capacity of trout Hb were at least partly due to low haemin affinity. Trout and tilapia Hb were equivalent in their ability to cause lipid oxidation in washed cod muscle heated to 80 ℃. Apparently, these high temp. denature both trout and tilapia Hb to such an extent[4] that any differences in conformational stability observed at lower temperature were negated.

2.词或词组

acid-inducing unfolding 酸诱发的解折叠
autoxidize 自动氧化
fatty acid unsaturation 脂肪酸不饱和度
haemoglobin (Hb) *n.* 血红蛋白
lipid oxidation 脂肪氧化
MetHb(= metmyoglubin) 高铁肌红蛋白
polyenoic *adj.* 多烯的
phospholipid hydrolysis 磷脂水解
secondary and tertiary structure 二级和三级结构
thermal aggregation 热聚合

3.主要语法现象

[1] as 引导的比较状语从句
[2] which 引导的非限制性定语从句

[3] 过去分词作定语

[4] such...that... 引导的结果状语从句

4.原文的参考译文

第5课　鱼肉血红素蛋白结构和脂肪组成对血红蛋白调节的脂肪氧化的影响

尽管经清洗的鳕鱼中多烯化合物的含量较经清洗的罗非鱼高2.8倍,但罗非鱼肌肉中血红蛋白(Hb)能更有效地促进脂肪氧化。这说明提高鱼肉中不饱和脂肪酸的含量并不会加速脂肪的氧化。而且观察到经清洗的鳕鱼中有相当含量的磷脂会被水解,而这具有抑制脂肪氧化的潜力。高铁血红蛋白(MetHb)的形成和脂肪氧化在pH 6.3较pH 7.4时更快一些。与罗非鱼相比,鳟鱼血红蛋白(Hb)自动氧化速率更快,且更快促进脂肪的氧化。鳟鱼血红蛋白促进脂肪氧化的能力更强,部分源于其较低的二级和三级构象及结构的稳定性、酸诱发的解折叠以及热聚合。该结果表明,鳟鱼血红蛋白结构的不稳定性和脂肪氧化的能力至少部分可归因于其较低的血红蛋白亲和性。当经清洗的鳕鱼肌肉加热到80 ℃时,鳟鱼和罗非鱼血红蛋白导致脂肪氧化的能力相当。显然,鳟鱼和罗非鱼血红蛋白在低温条件下观察到的任何结构稳定性差异程度,高温变性作用均可被忽略。

第6课　Comparative whole-grain intake of British adults in 1986—1987 and 2000—2001

1.原文

引自: Food Science and Technology Abstracts,2007,39(9):204

源自: British Journal of Nutrition,2007,97(5):987-992

Epidemiological evidence suggests that higher consumption of whole-grain food can significantly reduce the risk of chronic diseases such as cardiovascular disease, type 2 diabetes and some cancers. The study compared whole-grain intake of 2 086 UK adults aged 16-64 years from the 1986—1987 Dietary and Nutritional Survey of British Adults with that of 1 692 adults aged 19-64 years from the 2000—2001 National Diet and Nutrition Survey. For each survey, whole-grain intake was estimated from the consumption of all foods containing = 10% whole-grain content (as DM/fresh wt. of food) from 7 day weighed dietary records. In 1986—1987, median whole-grain intake was 16 (interquartile range 0.45) g/day vs. 14 (interquartile range 0.36) g/day in 2000—2001($P<0.001$). In 1986—1987, 77% of adults had less than three 16 g amounts of whole-grain intake/day; 25% reported no whole-grain intake. In 2000—2001, corresponding percentages were 84% and

29%, respectively. Foods with <51% whole-grain content provided 18% of whole-grain intake in 1986—1987 vs. 27% in 2000—2001 ($P<0.001$). In both surveys, whole-grain intake was significantly lower among adults with a manual vs. non-manual occupation (indicative of lower socioeconomic status) and among smokers vs. non-smokers, independent of occupational social class. In 1986—1987, whole-grain breakfast cereals and wholemeal bread contributed 28% and 48% of wholegrain intake, respectively, vs. 45% and 31% in 2000—2001. At each time, 1/3 of adults consumed neither of these 2 largest contributors to wholegrain intake. These findings suggest that wholegrain intake of British adults was low in 1986—1987 and became even lower over the subsequent decade.

2.词或词组

cardiovascular disease　心血管病

Dietary and Nutritional Survey of British Adults　英国成年人膳食与营养普查

epidemiological evidence　流行病学证据

National Diet and Nutrition Survey　全国膳食与营养普查

type 2 diabetes　Ⅱ型糖尿病

whole-grain intake　全谷物粗粮摄入

3.原文的参考译文

第 6 课　1986—1987 年度和 2000—2001 年度英国成年人全谷物摄入量比较

流行病学证据显示多吃全谷物可以显著减少患慢性病如心血管病、Ⅱ型糖尿病和某些癌症的危险性。本研究将 1986—1987 年度英国成年人膳食与营养普查时年龄为 16～64 岁的 2 086 位英国成年人的全谷物摄入量与 2000—2001 年度全国膳食与营养普查时年龄为 19～64 岁的 1 692 名的成年人的全谷物摄入量进行了比较。每次普查，全谷物的摄入量均按记录的 7 d 的定量膳食摄入总食物量的 10%（以干物质/食物鲜重）估计。1986—1987 年度和 2000—2001 年度的全谷物粗粮平均摄入量分别为 16 g/d(四分位差为 0.45)和 14 g/d(四分位差为 0.36)($P<0.001$)。在 1986—1987 年度，77% 的成年人每顿饭的全谷物摄入量少于 16 g，25%的人表示根本不摄入全谷物粗粮；2000—2001 年度，相应的比例分别为 84% 和 29%。在 1986—1987 年度，全谷物粗粮摄入量的 18%是由全谷物含量少于 51%的食物提供的，而 2000—2001 年度占 27%($P<0.001$)。在这 2 次普查中，成年体力劳动者全谷物粗粮摄入量明显低于非体力劳动者(社会经济地位较低者)，吸烟者明显低于不吸烟者(与职业和社会地位无关)。在 1986—1987 年度，全谷物早餐食物和全麦面包分别占全谷物粗粮摄入量的 28%和 48%，而 2000—2001 年度该比例则分别为 45%和 31%。每次普查中都有 1/3 的成年人不是通过上述 2 种主要的途径来摄入全谷物的粗粮。这些发现表明英国的成年人在 1986—1987 年度其全谷物粗粮摄入量是很低的，而且在随后的十多年中变得更低。

第7课 Effect of high-pressure treatment on survival of *Escherichia coli* O157:H7 population in tomato juice

1.原文

引自: Food Science and Technology Abstracts,2007,39(9):143

源自: Journal of Food, Agriculture & Environment,2007,5(1):111-115

Liquid foods have been implicated in numerous food-borne outbreaks and recalls. In this study, tomato juice and phosphate buffer saline[1] inoculated with *Escherichia coli* O157:H7 at 10^8 CFU/mL was exposed to continuous or repeated cycles of high-pressure (300 to 600 MPa) treatment at 25 ℃. Treatments using moderate pressure at 300, 350 and 400 MPa for up to 60 min reduced the population of inoculated *E. coli* O157: H7 by approximately 3.0, 3.0 and 5.0 log cycles, respectively in tomato juice. Population of *E. coli* O157: H7 in all samples decreased when temperature was increased from 30 to 40 ℃; however, increase in temperature and a 600 MPa pressure resulted in more sublethal injury on cells inoculated in phosphate buffer than tomato juice. Acidity of the tomato juice killed and prevented recovery of the injured bacteria. Repeating the pressure 4 times at 300 MPa for a total of 40 min at room temperature was significant ($P<0.05$) in reducing *E. coli* populations by an extra 1.0 log. Results indicated that repeating high-pressure treatment at least 4 times at 300 MPa for total of 40 min would inactivate more *E. coli* O157:H7 strains in tomato juice than using continuous pressure for the same amount of time.

2.词或词组

Escherichia coli O157:H7　大肠杆菌 O157:H7

food-borne outbreak　食源性疾病的暴发

inoculate　*v.*　接种

phosphate buffer saline　磷酸盐缓冲溶液

recall　*n.* 召回；*vt.*　回忆，回想，记起，取消

sublethal injury　亚致死损伤

3.主要语法现象

[1] 过去分词作定语

4.原文的参考译文

第 7 课　高压处理对番茄汁中大肠杆菌 O157:H7 存活的影响

多起食源性疾病的暴发和产品召回都与液体食品有关。在本研究中，将番茄汁和磷酸盐缓冲溶液接种大肠杆菌 O157:H7，使菌落数为 10^8 CFU/mL，然后在 25 ℃条件下分别利用 300～600 MPa 的压力连续或间歇性处理一定的时间。利用较低的 300、350 和 400 MPa 的压力处理番茄汁 60 min 可以使番茄汁中大肠杆菌 O157:H7 数分别降低 3.0、3.0 和 5.0 个对数级。当将温度由 30 ℃提高到 40 ℃，则所有样本中大肠杆菌 O157:H7 的数量均会减少。然而，与番茄汁处理组相比，提高温度和用 600 MPa 的压力处理，磷酸盐缓冲溶液组中有更多的大肠杆菌 O157:H7 受到亚致死损害。番茄汁的酸度会杀死和防止受损的细胞重新修复。室温条件下利用 300 MPa 的压力 4 次重复处理番茄汁 40 min 可以使大肠杆菌 O157:H7 的数量额外减少一个对数级（$P<0.05$）。这一结果显示利用 300 MPa 的压力 4 次重复处理番茄汁 40 min 可以比同样的压力条件下一次处理同样长的时间更有效地使更多的大肠杆菌 O157:H7 失活。

第 8 课　Improvement of the overall quality of the table grapes stored under modified atmosphere packaging in combination with natural antimicrobial compounds

1.原文

引自：Food Science and Technology Abstracts，2007，39(9)：167

源自：Journal of Food Science，2007，72(3)：S185-S190

Consumers demand new means of preservation with absence of chemicals. In this work, a package was developed (thermosealed baskets) with table grapes wrapped with 2 distinct films (M and P) with different permeability (medium and high, respectively) without or with the addition of a mixture of eugenol, thymol and carvacrol. Table grapes stored on air (control) lost their quality attributes very rapidly, manifested by accelerated weight loss, color changes, softening and increase in soluble solids concentration and titratable acidity ratio(SSC/TA). Use of modified atmosphere packaging (MAP) alone retarded these changes,[1] the effects being significantly greater when essential oils were added (especially for M film), although atmospheric composition was not affected by incorporating essential oils. In addition, microbial counts (fungi, yeasts and mesophilic aerobes) were decreased markedly and accompanied by a lower occurrence of berry decay.

Although slight aroma was detected after opening the packages, absence of the typical flavor of these compounds was found by trained panelists after tasting the grapes. Results suggest that the overall quality (sensory and safety) of table grapes could be improved and[2] the method considered an alternative to the use of synthetic fungicides.

2.词或词组

carvacrol *n*. 香芹酚
essential oil 精油
eugenol *n*. 丁香油酚
fungicide *n*. 防霉剂
modified atmosphere packaging (MAP) 气调包装
mesophilic aerobe 嗜温性好氧菌
microbial count 微生物的数量
permeability *n*. 通透性
soluble solids concentration 可溶性固形物浓度
thermoseal *vt*. 热封
thymol *n*. 百里酚,麝香草酚
titratable acidity 可滴定酸度
trained panelist 接受过培训的品尝人员

3.主要语法现象

[1] 独立分词结构
[2] 宾语从句

4.原文的参考译文

第8课 气调包装结合天然抗菌剂能有效改善鲜食葡萄在贮藏中的整体品质

消费者要求用非化学的新方法来保藏食品。本研究中,开发了一种包装(采用热封),对鲜食葡萄用2种(M和P)通透性不同(分别为中等和高通透性)的膜进行包装,并在包装袋中分别添加或不添加由丁香油酚、百里酚及香芹酚组成的混合物。对照组中(大气条件下)鲜食葡萄的品质迅速丧失,伴随着重量损失加速、颜色变化、组织变软及可溶性固形物(SSC)与可滴定酸(TA)的比值(SSC/TA)升高。气调包装(MAP)则能有效阻止这些变化,当添加上述精油后效果更佳(尤其对用中等通透性膜包装的),尽管添加上述精油并不会改变包装中气体的组成。此外,微生物的数量(真菌、酵母和嗜温性好氧菌)会显著减少,果品腐烂较少。尽管在打开包装后可以感受到少量的芳香物质,但专业品尝人员在品尝了这些葡萄后并未觉察到这些化学物质的特征性风味减少。这些结果说明了上述处理可以改进鲜食葡萄的整体品质(包括感官品质和安全性),并且可作为使用合成防霉剂的替代方法。

第 9 课　Optimization of an enzyme assisted process for juice extraction and clarification from litchis (*Litchi chinensis* Sonn.)

1. 原文

引自：Food Science and Technology Abstracts, 2007, 39(9): 140

源自：International Journal of Food Engineering, 2007, 3(2)

Litchi pulp was treated with various concentration of hydrolytic enzymes, including pectinase (polygalacturonase; 0-0.133% *W/W*), cellulase (0-0.266% *W/W*) and hemicellulase (0-0.20% *W/W*) for different durations (30-150 min) at 45 ℃. Effects of enzyme treatment conditions were studied on yield, clarity, apparent viscosity and TSS contents of juice obtained from the pulp. The optimum process conditions were determined by employing a 2nd order central composite rotatable design in combination with response surface methodology. Yield, clarity and TSS of juice were found to increase and apparent viscosity was found to decrease significantly after enzymic treatment. The optimum conditions for max. juice yield, clarity and TSS and min. viscosity were 0.076% (*W/W*) pectinase, 0.138% (*W/W*) cellulose and 0.107 (*W/W*) hemicellulase; with an incubation time of 106.5 min. The predicted value for juice yield, clarity, viscosity and TSS under the optimized conditions were 77.19%, 93.53%, 1.359 MPa · s and 19.68 °Brix, respectively, which showed a good agreement with experimental values under the same set of conditions.

2. 词或词组

apparent viscosity　表观黏度

°Brix　*n.*　糖度

cellulose　*n.*　纤维素酶

hydrolytic enzyme　水解酶

hemicellulase　*n.*　半纤维素酶

2nd order central composite rotatable design　二阶中心复合旋转设计

pectinase　*n.*　果胶酶

polygalacturonase　*n.*　多聚半乳糖醛酸酶

response surface methodology　响应面法

TSS = total soluble solid　总可溶性固形物

3.原文的参考译文

第9课　在荔枝(*Litchi chinensis* Sonn.)汁的榨取和澄清中酶法反应条件的优化

运用各种不同浓度,不同类型的水解酶包括果胶酶(多聚半乳糖醛酸酶;0～0.133%,质量分数)、纤维素酶(0～0.266%,质量分数)及半纤维素酶(0～0.200%,质量分数)在45℃条件下处理荔枝果浆30～150 min,研究酶处理的条件对果汁产品出汁率、澄清度、表观黏度和总可溶性固形物(TSS)含量的影响。采用二阶中心复合旋转设计结合响应面法确定最佳的反应条件。酶处理后发现果汁产品的出汁率、澄清度和总可溶性固形物(TSS)含量会显著增加而表观黏度会显著下降。获得最大的产品的出汁率、澄清度和总可溶性固形物(TSS)含量及最小黏度的最佳酶反应条件为果胶酶浓度为0.076%(质量分数)、纤维素酶浓度为0.138%(质量分数)及半纤维素酶浓度为0.107%(质量分数),作用时间为106.5 min。在此条件下,出汁率、澄清度、表观黏度和总可溶性固形物(TSS)含量的预测值分别为77.19%、93.53%、1.359 MPa·s和19.68 °Brix,这与同等条件下验证实验所获得的结果一致。

作　业　汉译英

(1)回生抗性淀粉种类对米淀粉凝胶形成的影响(谢涛,李英,易翠平,等.农业工程学报,2017,33(4):309-314)

摘要:为寻找改善普通米淀粉制品的结构及品质的新型食品添加剂,该文以普通米淀粉为原料,采用快速黏度分析仪、扫描电子显微镜、质构分析仪、全自动X射线衍射仪及示差扫描量热仪等手段,研究添加锥栗、马铃薯和绿豆回生抗性淀粉(retrograded resistant starch,RSⅢ)对米淀粉凝胶微观结构及理化性质的影响。结果表明:添加锥栗、马铃薯及绿豆RSⅢ对米淀粉凝胶的结构及性质产生显著影响($P<0.01$),以锥栗RSⅢ的作用最为突出。添加锥栗、马铃薯与绿豆RSⅢ对米淀粉糊的黏度特性没有影响($P>0.05$)。未添加RSⅢ的米淀粉凝胶存在很多不规则、深浅不一的大洞,而加入RSⅢ使米淀粉凝胶的网状结构变得更为规整、致密,且其胶着性与黏聚性变化不大($P>0.05$);添加锥栗、马铃薯与绿豆RSⅢ后能加速米淀粉凝胶的形成,与未添加RSⅢ的米淀粉凝胶比,其硬度分别增加了2.38、1.97和1.25倍($P<0.01$),黏着性分别增加2.56、1.99和1.32倍($P<0.01$),弹性增加1.07、0.81和0.53倍($P<0.01$)。米淀粉以A-型晶体占优,锥栗RSⅢ以V-型晶体占优,马铃薯与绿豆RSⅢ均以B-型晶体占优;不加或加入RSⅢ的米淀粉凝胶粉末都转变为以V-型晶体为主,且总相对结晶度没有改变($P>0.05$)。加入RSⅢ后的米淀粉糊除有低温吸热峰外还出现高温吸热峰,是否添加RSⅢ对低温吸热峰的温度参数影响不大($P>0.05$),但吸热焓显著降低($P<0.01$);而对于高温吸热峰,添加马铃薯与绿豆RSⅢ的各项参数没有差别($P>0.05$),但比添加锥栗RSⅢ的显著增高($P<0.01$)。可见添加不同来源的RSⅢ可以有效改善米淀粉凝胶的结构与品质。该研究结果为抗性淀粉用于提高米制品品质与营养功能的研究和生产提供

了重要参考。

(2)冰温保鲜对牛肉肌原纤维蛋白结构和功能特性的影响(夏秀芳，李芳菲，王博，等. 中国食品学报,2015,15(9):54-60)

摘要:以冷藏(4 ℃)的牛肉为对照,通过测定牛肉肌原纤维蛋白的 ATP 酶活性、巯基含量、溶解性、热稳定性和电泳等指标,研究冰温(−1 ℃)保鲜对牛肉蛋白质结构和功能特性的影响。结果表明:随着贮藏时间的延长,牛肉肌原纤维蛋白的功能性显著降低。SDS-PAGE 表明蛋白质发生不同程度的降解。在相同贮藏时间内,冰温保鲜比冷藏条件可减少蛋白质结构和功能性的变化。贮藏至第 12 天时,4 ℃冷藏牛肉肌原纤维蛋白的 Ca^{2+}-ATPase 活性下降了 72.9%,而冰温保鲜的牛肉下降 46.9%;冷藏牛肉肌原纤维蛋白的总巯基和活性巯基含量分别下降 50.12%和 93.68%,而冰温保鲜牛肉降至 35.08%和 71.15%;冷藏牛肉蛋白质溶解度下降 41.18%,而冰温保鲜牛肉下降 15.52%。SDS-PAGE 表明冷藏牛肉的肌原纤维蛋白降解程度比冰温保鲜的大,冷藏条件下的牛肉肌原纤维蛋白热稳定性也不如冰温保鲜的高。冰温保鲜是一种有效的牛肉保鲜方法。

(3)聚合乳清浓缩蛋白对发酵乳饮料稳定性的影响(孙亚婷,蒋姗姗,曹宋宋,等. 中国食品学报,2018,18(3):157-163)

摘要: 以发酵乳饮料为研究对象,探讨聚合乳清浓缩蛋白的添加量对发酵乳饮料稳定性的影响。考察了聚合乳清浓缩蛋白替代总蛋白对发酵乳饮料的离心沉降率、吸光度比值、粒径、黏度、表面疏水性、游离巯基的影响。研究结果表明:添加聚合乳清浓缩蛋白的发酵乳饮料离心沉降率显著低于未添加聚合乳清浓缩蛋白的发酵乳饮料($P<0.05$);随着聚合乳清浓缩蛋白替代总蛋白比例的增加,发酵乳饮料的吸光度比值先增加后略有减小;而发酵乳饮料的黏度逐渐增大;发酵乳饮料的粒径先减小后略有增大,且聚合乳清浓缩蛋白替代总蛋白比例为30%时,体系粒径最小;发酵乳饮料的疏水性和游离巯基逐渐增大,聚合乳清浓缩蛋白替代总蛋白比例为 10%～40%的发酵乳饮料的游离巯基显著大于空白($P<0.05$)。研究结果表明,适当添加聚合乳清浓缩蛋白有利于发酵乳饮料的稳定。

(4)苦荞对高脂膳食诱导小鼠生理及肠道菌群的影响(泰驿,闫贝贝,王宏,等. 食品科学,2018,39(1):172-177)

摘要: 以苦荞为研究对象,通过高脂膳食诱导建立血脂代谢紊乱小鼠模型,利用高效液相色谱和平板计数对血脂代谢指标和肠道菌群的变化规律及其相关性进行研究,初步探究了苦荞对血脂代谢紊乱小鼠生理及肠道菌群的影响。结果表明:高脂膳食可引发小鼠血脂代谢紊乱,苦荞淀粉和蛋白均能显著降低血脂指标(血清总胆固醇、甘油三酯、低密度胆固醇)的水平($P<0.05$),并基本恢复至空白组的水平。通过平板计数对各组小鼠的肠道主要菌群(双歧杆菌、乳酸菌、大肠杆菌、肠球菌)检测,发现苦荞淀粉和苦荞蛋白干预组的小鼠肠道菌中有益菌(双歧杆菌、乳酸菌、肠球菌)数量均显著高于高脂组($P<0.05$),有害菌(大肠杆菌)数量则显著低于高脂组($P<0.05$)。此外,相关性分析结果表明,苦荞中蛋白和淀粉均可吸附胆汁酸与胆固醇并促进其排泄,从而不仅能调节血脂代谢又能调节肠道菌群平衡,同时,肠道菌群中益生菌比例的增多和有害菌比例的降低又对血脂代谢调节与抑制氧化应激有着一定促进作用。因此,苦荞对血脂代谢调节作用可能与其促进胆酸排泄、调节肠道菌群平衡、改善氧化应激等方面有着密切关系。

第 2 章

综　述

Review

第 1 课　Hazards of packaging materials in contact with foods

1. 原文

选自： Trends in Food Science & Technology，2007，18(4)：219-230

作者： Poças M F，Hogg T

Synthetic polymers typically have high molecular weights (5 000-1 000 000 million u) and therefore their biological availability is negligible. However,[1] due to the use of lower molecular weight (<1 000 u) additives in these polymers as well as the presence of trace levels of unreacted monomers, there is a finite potential for human exposure to these lower molecular weight components. Substances[2] that may migrate from plastic materials include monomers and starting substances, catalysts, solvents and additives. This latter class includes antioxidants, antistatics, antifogging agents, slip additives, plasticizers, heat stabilisers, dyes and pigments.

Paper and board are essentially composed of pulp from different vegetable sources and are most often employed in contact with dry foods. Additives[3] used in this type of material include fillers, starch and derivatives, wet strength sizing agents, retention aids, biocides, fluorescent whitening agents and grease-proofing agents. Paper and board may also be coated with polymers as polyethylene or waxes. Recycled fibre is considered a major source of migrants. This route of contamination is officially recognized in the Resolution RESAP (2002)1 of the Council of Europe for paper and board in contact with foods,[4] which lists DIPNs, benzophenone, partially hydrogenated terphenyls, solvents, phthalates, azo-colourants, primary aromatic amines and polycyclic aromatic hydrocarbons as being relevant. Corrugated board is most often used as transport packaging system and thus not anticipated to come into direct contact with food. However, volatile substances in this type of material[5] used as a secondary package may be transferred through the primary packaging into the food.

Metal cans are made of tin-plate (steel coated with tin). Tin-free steel (steel coated with chromium and chromium oxides) or aluminium. Tin-plate is most used for food cans and aluminium for beverage cans. Most cans are internally coated with a polymeric layer, and thus the layer of food contact is not the metal but the lacquer. The substances of concern in can systems are therefore not only the metals involved, but also components migrating from the coatings, such as starting substances and their potential derivatives. Migrants from can coatings, namely phenolic resins, often contain only small amounts of monomers, oligomers and additives, but a large amount of other unknown or undescribed components.

Glass packaging has as its major components, silica, sodium and calcium oxides. These components are unlikely to have any significant effect on the safety of foods since they are natural constituents of many foods. Silica is also the major component of food-contact ceramics. Clays, another major raw material of ceramics, is composed of alumina, silica and water. Substances of concern may, however, originate from glazes and printing inks. Thus lead and cadmium are frequently controlled in such materials since they may be present as contaminants. The Food Standards Agency (UK) promoted a comprehensive overview of the potential for elemental migration from different glass types[6] used in food-contact applications in a range of conditions of use.

2.词或词组

antistatic *adj*. 抗静电的

retention *n*. 保持力

corrugate *v*. 弄皱,起皱,(呈)波纹状

glaze *n*. 釉料、釉面,光滑面; *v*. 上釉,使表面光滑

phthalate *n*. 邻苯二甲酸盐

azo-colourant *n*. 偶氮着色剂

3.主要语法现象

[1] 状语从句

[2][4] 定语从句

[3][5][6] 过去分词作定语,放在被修饰词之后

4.原文的参考译文

第1课 与食品接触的包装材料的危害性

合成聚合物具有很大的分子质量(5 000～1 000 000 Mu),因此,难以被生物利用。然而,由于在这些聚合物中使用了低分子质量(<1 000 u)的添加剂以及一些痕量的未反应的单体,人体摄入这些低分子质量化合物,会受到潜在的危害。可能从塑料材料中转移出的物质有单体和引物、催化剂、溶剂和添加物等。添加物包括抗氧化剂、抗静电剂、防雾剂、润滑剂、可塑剂、热稳定剂、染料和色素等。

纸和纸板实质上都是由各种植物来源的纸浆沉积而成的,是用于包装干食品最主要的材料。在这些材料中使用的添加物有填充物、淀粉和淀粉衍生物、湿强剂、保存剂、防虫剂、荧光剂、漂白剂和阻油剂等。纸盒纸板还可用聚乙烯或蜡等聚合物涂层处理。再生纤维被认为是主要的污染来源。欧洲直接接触食品用纸和纸板协会制定的RESAP(2002)1决议中认定了纸和纸板存在的污染途径,其中列出的相关污染物包括:DIPNs、苯甲酮、部分氢化三联苯、溶剂、邻苯二甲酸盐、含氮着色剂、初级芳香胺、多环芳烃等。瓦楞纸板是常用的承重包装材料,不可用于与食品直接接触的包装,然而,用作外包装的瓦楞纸板中的挥发性物质可能会透过与食品直接接触的内包装进入食品中。

金属罐由镀锡钢(钢表面涂锡)、无锡钢(表面镀铬和铬氧化)或铝材料制成。锡罐主要用

于食品罐头，铝罐主要用于饮料罐头。许多罐子的内表面涂有聚合物，因此，与罐内食品直接接触的是涂漆而不是金属。因此，罐头体系中相关的污染来源不仅有金属材料，还有涂料渗出物，如涂料及其反应产物。罐涂层渗出物，即酚醛树脂，常含少量的单体、低聚体和添加剂，还含有大量其他未知的或未确定的物质。

包装用玻璃材料中主要成分是硅、钠和钙的氧化物。这些成分对食品安全不可能构成明显的影响，因为它们也是许多食品中含有的天然成分。硅土也是用于盛装食品的陶瓷的主要成分，另一种生产陶瓷的主要原料——陶土，它是由氧化铝、硅土和水组成。然而需注意来源于釉彩和印染油墨的污染。铅和镉是这些材料中的主要污染物，应严格控制。英国食品标准机构较全面地分析了各种条件下用于食品包装的各种玻璃材料的潜在污染物。

第2课 Iron deficiency and iron fortified foods

1.原文

选自： Food Research International，2002，35(2-3)：225-231

作者： Martínez-Navarrete N，Camacho M M，Martínez-Lahuerta J，Martínez-Monzóand J，Fito P

Iron deficiency is considered to be the commonest worldwide nutritional deficiency and affects approximately 20% of the world population. Women and young children are especially at risk. It is stated[1] that adverse effects include lower growth rate and impaired cognitive scores in children and poor pregnancy outcome and lower working capacity in adults. This deficiency is partly induced by plant-based diets, such as vegetarian diets, containing low levels of poorly bioavailable iron. In this sense, it is as widespread in less industrialized countries as in developing countries. Several works show a prevalence of nutritional anemia of around 20%-50% in the former and between 2% and 28% in developed countries. The global nature of the problem and its public health significance has been reviewed by Hambraeus (1999).

In addition, iron demand may be increased by blood loss and by fast growth,[2] which increases the risk of iron deficiency in infants, young adolescents, and in menstruating and pregnant women. As an example, the study reported by Arija et al. (1997) shows data of iron deficit (serum ferritin alteration) in Spanish population grouped by age. Prevalence[3] observed by these authors was 19% in children from 0.5 to 3 years old, 14% for 4-8 year olds, 10% for 8-12 year olds, 11% for 12-16 year olds, 1% for males between 18 and 65 years old, 14% for females between 18 and 54 years old and 1% for females between 54 and 65 years old. Nevertheless, genetically determined diseases may lead to siderosis, such as hereditary haemochromatosis or thalassaemia major, although these cases show a limited geographic and ethnic distribution.

[4] Recommended daily intakes of dietary iron for normal infants are 1 mg iron per kg per day and for children and male and female adolescents, 10, 12 and 15 mg per day. For

women during reproductive years, 15 mg per day and adult men and postmenopausal women require only 10 mg per day.

Prerequisites for effective supplementation include an efficient and consistent supply, delivery, and consumption of a highly bioavailable iron supplement. To be effective, a combination of an iron fortificant compound and food vehicle must be selected[5] which is safe, acceptable to and consumed by the target population, does not adversely affect the organoleptic qualities and shelf-life of the food vehicle, and provides iron in a stable, highly bioavailable form. Bioavailability of both the fortificant and the intrinsic food iron can be improved by adding enhancing factors, removing inhibitors such as phytate by enzymatic and non-enzymatic hydrolysis, and using 'protected' fortification compounds. Factors such as pH are also important in determining iron bioavailability. The higher the pH, the greater the amount of ferrous ions oxidized to the ferric state. The presence of about 10% ferric ions results in discoloration.

2.词或词组

bioavailable *adj*. 生物可利用的

ferritin *n*. 铁蛋白

fortificant *adj*. 强化的;增强的

haemochromatosis *n*. 血色素沉着症

menstruating *n*. 月经,月经期间

organoleptic *adj*. 感官感觉的,能接受感觉的

siderosis *n*. 肺铁末沉着病:由于过度吸入金属铁尘或氧化铁的粉尘而引起的慢性肺部炎症

thalassaemia *n*. 地中海贫血(症)

3.主要语法现象

[1][5] 主语从句

[2] 非限制性定语从句

[3] 分词短语作定语

[4] 分词短语作主语

4.原文的参考译文

第2课 铁缺乏及铁强化食品

铁缺乏是最普遍的世界性营养缺乏病,全世界大约20%的人口受其影响,尤其是妇女和儿童。铁缺乏可能导致儿童生长缓慢、智力低下,成人怀孕率低和工作效率差。铁缺乏部分是由大量食用生物活性铁含量低的植物性食物引起的,如素食。这种膳食在工业化程度低的国家和发展中国家都很普遍。一些研究表明在这些国家中营养性贫血占20%~50%,而在发达国家只占2%~28%。Hambraeus 等曾对这一问题的世界状况及其对公众健康的重要影响进

行过论述。

另外,失血和身体快速生长都可能增加人体对铁的需求量,因此,婴幼儿、青少年及处于月经期和怀孕期的妇女都会增加缺铁的风险。例如,Arija 等报道西班牙各年龄段的缺铁人员的比例分别为:0.5~3 岁婴幼儿中占 19%,4~8 岁儿童中占 14%,8~12 岁儿童中占 10%,12~16 岁青少年中占 11%,18~65 岁男性中占 1%,18~54 岁女性中占 14%,54~65 岁女性中占 1%。然而,遗传性疾病也可引起肺铁末沉着病,如遗传性地中海贫血症和血色素沉着症,虽然这些病例局限于某些地区和种族。

铁摄入量的推荐值分别是:正常婴幼儿 1 mg/(kg · d);儿童、男青少年、女青少年分别为 10、12、15 mg/d;生育期女性 15 mg/d;成年男性和绝经期女性只需 10 mg/d。有效补充铁的前提条件包括高效铁补充剂的有效合理的提供、流通和消费。所谓铁强化剂的效力是指:铁强化剂和食物载体间的结合必须是安全的、易被消费人群接受和食用,且不影响产品的感观品质和保质期,并提供稳定的、生物效价高的产品形式。通过添加增效剂、脱除抑制剂(如用酶或非酶水解方法脱除植酸)和使用包埋保护等方法来提高和改善强化铁和食品中本身含有的铁的生物利用率。pH 等因素对铁的生物利用率也很重要,pH 越高,二价铁越易被氧化成三价铁。约 10%的三价铁存在就会引起产品变色。

第 3 课 Enzymatic deacidification/reesterification

1. 原文

选自: Journal of Food Engineering, 2005, 69 (4): 481-494

作者: Bhosle B M, Subramanian R

In this biorefining method, the unique ability of some microbial lipases to synthesize a triglyceride from a fatty acid and glycerol has been exploited to develop an alternative process for deacidifying vegetable oils with high-FFA contents. In view of the need for low-energy processes, microbial lipase-catalyzed esterification appears to be more advantageous for deacidification than chemical esterification,[1] which is invariably carried out at higher temperatures (180-200 ℃) than the lipase-catalyzed reactions. The microbial lipase process is also promising in terms of final quality of refined oils and refining loss. The potential of an enzymatic deacidification process for refining depends on several enzymatic esterification reaction variables, such as enzyme concentration, reaction temperature, reaction time, glycerol concentration, amount of moisture in the reaction mixture, pressure employed etc. Enzymatic deacidification of different vegetable oils has already been achieved on a laboratory scale. Sengupta and Bhattacharyya (1989) successfully brought down the FFA content of RBO from 30% to 3.6% by esterification of the FFA with added glycerol, using a 1,3-specific lipase (*Mucor miehei*). This process produced a RBO of excellent quality by subsequent alkali deacidification, bleaching, and deodourization. The [2] combined biorefining and alkali refining process compared well in terms of refining factor and colour with the miscella refining process. And with regard to refining characteristics, it was even superior to the combined physical refining and alkali neutralization process. Sengupta and Bhattacharyya (1992) showed that the FFA of mohua oil (*Madhuca latifolia*) could be reduced from 24.5% to a level of 3.8%, when the degummed and bleached oil was treated continuously with 10% lipase (*M. miehei*) and the stoichiometric amount of glycerol for 20 h at a temperature of 60 ℃ and a pressure of 267 Pa. Makasci, Arisoy, and Telefoncu (1996) examined the potential of enzymic deacidification for degummed and dewaxed hyperacidic olive oil. The lipase used was produced by a host *Aspergillus oryzae* from a selected strain of *Candida* and immobilized on a macroporous acrylic resin. They showed[3] that maintaining low pressure (or bubbling dry nitrogen) was important for the removal of water[4] formed during esterification.

The main advantage of the esterification process for deacidifying vegetable oils with high-FFA contents is the increase in the content of neutral glycerides, especially TG. However, the chief barrier to enzymatic deacidification is the high cost of enzymes.

2.词或词组

deacidification *n.* 脱酸
deodourization *n.* 脱臭
esterification *n.* 酯化
RBO (rice bran oil) *n.* 米糠油
macroporous *adj.* 大孔性的
acrylic *adj.* 丙烯酸的
resin *n.* 树脂
stoichiometric *adj.* 化学计量的

3.主要语法现象

[1] 定语从句
[2] 分词短语作定语,放在被修饰词之前
[3] 宾语从句
[4] 分词短语作定语,放在被修饰词之后

4.原文的参考译文

第3课 酶法脱酸和再酯化

在生物精炼方法中,游离脂肪酸含量高的植物油的脱酸工艺中应用了一些微生物脂肪酶具有将脂肪酸和甘油合成甘油三酯的特有功能。考虑到低能耗加工的需求,微生物脂肪酶催化酯化脱酸比化学法酯化更具优势,因为相对于脂肪酶催化反应,化学法酯化需要更高的反应温度,通常在180~200 ℃。微生物脂肪酶法工艺在提高精炼油成品品质和降低精炼损失方面也更有前途。酶脱酸工艺在油脂精炼上的潜力取决于酶酯化反应的几个变量,如酶浓度、反应温度、反应时间、甘油浓度、反应混合物中水分含量、应用的压力等。几种植物油的酶脱酸技术

已在实验室成功获得。Sengupta 和 Bhattacharyya(1989)用1,3-专一性脂肪酶(*Mucor miehei*)和甘油将FFA酯化,成功地将米糠油(RBO)中游离脂肪酸的浓度从30%降到3.6%。经后期的碱脱酸、脱色和脱臭处理,这种工艺可生产高品质的米糠油(RBO)。生物精炼和碱精炼相结合的生产工艺,对精炼因素和油的色泽而言,比油水精炼工艺更好;而对精炼的品质,它甚至优于物理精炼与碱中和相结合的工艺。Sengupta 和 Bhattacharyya (1992)报道脱胶和漂白处理的宽叶紫荆木油用10%脂肪酶和化学当量的甘油,在温度60 ℃、压力267 Pa的条件下,连续处理20 h,可将油中FFA含量从24.5%降到3.8%。Makasci, Arisoy 和 Telefoncu (1996)也试验了用酶脱酸法精炼经脱胶和脱蜡处理的高酸度橄榄油的可能性,采用的脂肪酶是从假丝酵母菌株中选择培养的 *Aspergillus oryzae* 中提取出来的,并固定在大孔丙烯酸树脂上。研究表明保持低压(或干燥氮气保护)对除去酯化过程中形成的水分很重要。

酯化法对高浓度FFA含量植物油进行脱酸处理的主要优势在于能增加油中中性甘油酯的含量,尤其是甘油三酯的含量。然而,酶的价格是酶法脱酸的主要障碍。

第4课 Physico-chemical properties of modified starch

1.原文

选自: Food Hydrocolloids,2007,21:1-22

作者: Singh J, Kaur L, McCarthy O J

The physico-chemical properties of starches such as swelling, solubility, and light transmittance have been reported to be affected significantly by chemical modification. The change in these properties upon modification depends on the type of chemical modification. Chemical modifications such as acetylation and hydroxypropylation increase, while cross-linking has been observed to decrease (depending on the type of cross-linking agent and degree of cross-linking) the swelling power and solubility of starches from various sources. The introduction of (bulky) acetyl groups into starch molecules by acetylation leads to structural reorganization owing to steric hindrance; this results in repulsion between starch molecules, thus[1] facilitating an increase in water percolation within the amorphous regions of granules and a consequent increase in swelling capacity. Studies[2] conducted on acetylated maize, potato and rice starches suggest a significant increase in swelling power and solubility upon acetylation in all these starch types. The extent of this increase was observed to be higher for potato starches. The degree of substitution[3] introduced after acetylation mainly affects the intensity of change in the swelling power and solubility of starches. The differences in the granule size distribution, physicochemical composition and granule rigidity among the starches may also be responsible for the alteration in swelling power and solubility after acetylation. The structural disintegration probably weakens the starch granules after acetylation, and this enhances amylose[4] leaching from the granule, thus[5] increasing starch solubility. The extent of the increase in swelling power has been observed to be higher in starches with a low amylose content and small size granule population. Liu et al. (1999b) reported that the waxy starches show an increased swelling power upon acetylation because of the presence of mainly amylopectin with a more open structure than in non-waxy starch; this allows rapid water penetration, and increased swelling power and solubility.

2. 词或词组

acetylation *n.* 乙酰化作用
amorphous *adj.* 无定性的,无组织的
disintegration *n.* 瓦解,碎裂
hydroxypropylation *n.* 羟丙基化作用
percolation *n.* 过滤,渗透
rigidity *n.* 硬度

3. 主要语法现象

[1][5] 分词短语作结果状语
[2][3][4] 分词短语作定语,放在被修饰名词之后

4. 原文的参考译文

第 4 课 变性淀粉的理化性质

化学改性会明显地影响淀粉的物理化学特性,如溶胀性、溶解性和透光率。因改性引起的这些特性变化取决于化学改性的类型,如乙酰化和羟丙基化改性处理会增加不同来源的淀粉的溶胀能力,而交联则减低淀粉的溶胀能力和溶解度(取决于交联剂和交联度)。乙酰化处理后在淀粉分子中引入了(大量的)乙酰基团,由于产生空间障碍而引起结构重组,淀粉分子间的相互排斥促进水渗透到淀粉颗粒的非结晶区,从而增加淀粉的溶胀能力。对乙酰化玉米淀粉、马铃薯淀粉和大米淀粉的研究结果均表明,乙酰化处理能明显提高这些淀粉的溶胀能力和溶解性。乙酰取代度主要影响淀粉溶胀能力和溶解性的变化程度。淀粉中颗粒大小分布、理化组成和颗粒强度上的差异也会影响乙酰化对淀粉的溶胀能力和溶解性的改变程度。乙酰化处理可能使淀粉结构崩解,削弱淀粉颗粒强度,从而增加直链淀粉从颗粒中溶出,因而增加淀粉的溶解性。研究发现,直链淀粉含量低和粒度小的淀粉的溶胀能力增加程度较高。Liu 等(1996b)报道蜡质淀粉经乙酰化处理后溶胀能力增加了,原因是蜡质淀粉主要含支链淀粉,而支链淀粉比非蜡质淀粉有更多的分支,有利于水分的迅速渗透,增加了溶胀能力和溶解性。

第 5 课 Preservation of wines

1. 原文

选自: Food Control, 2008, 3

作者: García-Ruiz A, Bartolomé B, Martínez-Rodríguez A J, Pueyo E, Martín-Álvarez P J, Moreno-Arribas M V

Sulphur dioxide (SO_2) has numerous properties as a preservative in wines, these include its antioxidant and selective antimicrobial effects, especially against lactic acid bacteria. Today, this is, therefore, considered to be an essential treatment in wine-making. However, the use of this additive is strictly controlled, since high doses can cause organoleptic alterations in the final product (undesirable aromas of the sulphurous gas, or when this is reduced to hydrosulphate and mercaptanes) and, especially, owing to the risks to human health of consuming this substance.

Because of these effects, in recent years there is a growing tendency to reduce the maximum limits[1] permitted in musts and wines. Although as yet, there is no known compound[2] that can replace SO_2 with all its enological properties, there is great interest in the search for other preservatives, harmless to health,[3] that can replace or at least complement the action of SO_2,[4] making it possible to reduce its levels in wines.

With regards products with antimicrobial activity complementary to SO_2, recently dimethyldicarbonate (DMDC) has been described as being able to inhibit alcoholic fermentation and development of yeasts,[5] permitting the dose of SO_2 to be reduced in some types of wines.

Yeast cells have been shown to die after adding this compound, whereas with SO_2 they enter a "viable state but cannot be cultivated",[6] which has also been demonstrated for lactic acid bacteria. Other alternatives have been introduced based on "natural antimicrobial agents",[7] of which the use of lysozyme is especially important, and some antimicrobial peptides or bacteriocins. In the case of lysozyme, since this was first authorized as an additive in wine-making it has only been used very little due to the high costs of its application. Another aspect to take into account about this protein is[8] that it can cause IgE-mediated immune reactions in some individuals so its presence in food products, including wine, can cause some concern. To date, nisin is the only bacteriocin[9] that can be obtained commercially, and although this has been shown to be effective at inhibiting the growth of spoilage bacteria in wines, it has not been authorized for use in enology. Other bacteriocins have been described to control the growth of lactic acid bacteria in wine, although the efficacy of these compounds, their mode of action and, especially, their stability during wine-making are still under investigation.

2. 词或词组

bacteriocin *n.* 细菌素

dimethyldicarbonate *n.* 二甲基碳酸氢钠

hydrosuplhate *n.* 硫化氢

immune *n.* 免疫者

lysozyme *n.* 溶菌酶

mercaptan *n.* 硫醇

3. 主要语法现象

[1] 分词短语作定语，放在被修饰名词后面

[2][3][4][6][7][9] 定语从句

[5] 分词短语,作补语

[8] 宾语从句

4.原文的参考译文

第5课 葡萄酒的保存

二氧化硫(SO_2)用于葡萄酒的防腐,具有许多特性:抗氧化性和选择性杀菌效果,尤其对乳酸菌。因此,它被认为是目前葡萄酒生产中基本的处理过程。然而,这种添加剂的使用应严格控制,高剂量 SO_2 会改变成品感官品质(硫黄气体或分解生成的硫化氢和硫醇的不愉快气味),尤其这种物质会危害消费者身体健康。

由于这些缺陷,近年来,降低 SO_2 在葡萄汁和葡萄酒中的最大限量已成为发展趋势。虽然如此,能完全替代 SO_2 在葡萄酒生产工艺中的作用的化合物还没被发现,研究其他的对人体无害的、能替代或至少能增加 SO_2 作用效果,以降低 SO_2 在葡萄酒中的用量的防腐剂,具有重要意义。

具有协同 SO_2 杀菌作用的化合物——二甲基碳酸氢钠能抑制酒精发酵和酵母的生长,可降低 SO_2 在葡萄酒中的添加量。当添加这种物质时,酵母细胞出现死亡,与 SO_2 合用能使酵母处于虽能存活但不能繁殖的状态。这对乳酸菌也有同样的效果。其他替代物源于天然杀菌剂和一些抗菌肽或细菌素,杀菌剂中溶菌酶的使用显得特别重要。溶菌酶虽然是最先允许在葡萄酒生产中使用的天然杀菌剂,但由于其使用费用较高,很少使用。值得关注的另一方面是,这种蛋白会使一些人产生 IgE 诱导的免疫反应,因此,含有这种蛋白的食品,如葡萄酒,会引起一些争议。目前,乳酸链球菌素是唯一能商品化生产的细菌素,已被证实对葡萄酒中败坏菌的生长有抑制效果,但尚未允许在葡萄酒工业中使用。其他的细菌素虽然能控制葡萄酒中乳酸菌的生长,但它们的作用机理和在葡萄酒生产中的稳定性还在研究中。

第6课 Effect of high pressure (HP) on microorganisms in milk

1.原文

选自:Trends in Food Science & Technology,2001,12(2):51-59

作者:O'Reilly C E, Kelly A L, Murphy P M, Beresford T P

As reported earlier, the pioneering research into the HP treatment of milk was carried out with the aim of developing an alternative to pasteurisation. Hite reported that the application of pressures of 1 400 or 460 MPa for 1 h preserved milk at room temperature for 4 days or 24 h, respectively. Hite also reported a 5-6 log reduction in microbial numbers when milk was treated at a pressure of 680 MPa for 10 min at room temperature.

The combined application of high hydrostatic pressure and heat treatment (67-71 ℃) to milk further extended shelf life. In 1965, Timson and Short found that the number of viable bacteria in milk was reduced to approximately one log cycle after pressurisation at 200 MPa for 30 min at 35 ℃. Later, Rademacher and Kessler concluded that, in order to achieve the shelf life of thermally pasteurised milk, 10 days at a storage temperature of 10 ℃, a pressure treatment of 400 MPa for 15 min or 500 MPa for 3 min was necessary.

Some recent studies have examined pressure inactivation of microorganisms[1] that are either naturally present in milk or introduced into the milk artificially. The decimal reduction values (D-value) of *Listeria monocytogenes* Scott A at 340 MPa in phosphate buffered saline, raw and UHT milk were 2.9, 9.3 and 13.2, respectively. These data indicate that milk was a protective medium against pressure inactivation for *L. monocytogenes* and that UHT treatment of the milk enhanced this affect. It has been suggested that this may be due to heat-labile antimicrobial compounds in raw milk[2] acting synergistically with pressure to enhance inactivation, or alternatively to the use of selective media[3] that inhibited the growth of sublethally injured cells. Erkmen and Karata HP-treated pasteurised whole cows milk (pH 6.4) inoculated with *Staphylococcus aureus* ATCC 27690 at pressures in the range 50-350 MPa for 4-12 min at 20 ℃. Extent of inactivation increased with pressure and treatment time. D-Values were estimated as 211.8, 15.0, 3.7 and 2.56 min at 200, 250, 300 and 350 MPa, respectively. Inactivation of *L. innocua* CECT 910 in ovine milk (6% fat) has been studied after treatment at 200-500 MPa for 0-60 min. A reduction in viable cell numbers of 6 and 5 log cycles were obtained on treatment at 450 MPa for 5 min at 2 and 25 ℃, respectively, while treatment at 300 MPa for 15 min at 50 ℃ resulted in a 7-log cycle reduction. The shelf life of whole and skim milk at 7 ℃ has been extended by pressure treatment at 400 MPa at 25 ℃ for 30 min.

It has been suggested that milk fat may have a baroprotective effect on pressure inactivation of microorganisms. However, such an effect was not observed by García-Risco et al., who suggested that the baroprotective effect might only become noticeable at fat levels higher than that of whole milk. Subsequent research with a range of microorganisms (*Escherichia coli*, *Pseudomonas fluorescens*, *L. innocua*, *S. aureus* and *Lactobacillus helveticus*)[4] inoculated into ovine milk, of varying fat content (0, 6 and 50%), indicated that the microorganisms were more resistant when treated at 100-500 MPa for 15 min at 4, 25 or 50 ℃ in 0% fat milk than in buffer. However,[5] varying fat levels did not appear to have any increased baroprotective effect, for any HP conditions or any microorganism tested.

2. 词或词组

baroprotective *adj*. 耐压的,压力保护的

hydrostatic *adj*. 静水力学的,流体静力学的

pasteurisation *n*. 巴氏杀菌法,加热杀菌法

pressurization *n*. 增压,耐压,加压

phosphate *n*. 磷酸盐

3.主要语法现象

[1][3] 定语从句

[2] 现在分词短语作定语

[4] 过去分词短语作定语

[5] 现在分词短语作主语

4.原文的参考译文

第6课 高压处理对牛奶中微生物的影响

较早的报道,高压处理牛奶最初的目的是替代巴氏杀菌。Hite 报道牛奶分别在 1 400 MPa 和 460 MPa 下处理 1 h 可相应地在室温下保存 4 d 和 24 h。他还报道在 680 MPa、室温条件下处理 10 min 可将牛奶中细菌总数下降 5～6 个数量级。采用高压处理与热处理(67～71 ℃)相结合的方式可进一步延长产品货架期。1965 年,Timson 和 Short 发现经 35 ℃、200 MPa 高压处理 30 min,牛奶中活菌总数下降到十位数,后来,他们得出结论,要达到巴氏热灭菌所获得的牛奶货架期,即 10 ℃ 条件下要保质 10 d,则必须在 400 MPa 下处理 15 min 或在 500 MPa 下处理 3 min。

最近的一些研究检测了压力处理对牛奶中天然的或加工过程中侵入的微生物的灭活效果。在 340 MPa 条件下,在磷酸盐缓冲液、原料奶和超高温灭菌奶中 *Listeria monocytogenes* Scott A 的对数死亡时间(*D* 值)分别为 2.9、9.3 和 13.2 min。这些数据表明牛奶是一种能提高 *L. monocytogenes* 的耐压性的介质,而超高温处理能强化这种作用。有人认为是由于牛奶中含有能与压力灭菌处理产生协同效应的热敏性抗菌成分,或者是由于使用了能抑制亚损伤细胞生长的专一性介质。Erkmen 和 Karata 研究了在 20 ℃、50～350 MPa 范围内高压处理接种有 *Staphylococcus aureus* ATCC 27690 的巴氏灭菌全脂牛奶 4～12 min 的灭菌效果。在 200、250、300、350 MPa 下 *D* 值分别为 211.8、15.0、3.7、2.56 min。也有人研究了含脂 6% 的羊奶在 200～500 MPa 下处理 0～60 min,*L. innocua* CECT 910 的失活情况,结果发现,在 450 MPa 2 ℃ 或 25 ℃ 条件下处理 5 min,奶中活菌数分别下降了 6 个和 5 个对数指数,而在 300 MPa 50 ℃ 条件下处理 15 min,下降了 7 个对数指数。400 MPa 25 ℃ 下处理 30 min 可延长全脂奶和脱脂奶在 7 ℃ 下的货架期。

有人认为奶中脂肪能提高微生物对压力处理的抵抗性。然而,García-Risco 等发现只在比全脂奶脂肪含量高的奶中出现明显的压力防护效果。在对含脂量分别为 0、6% 和 50% 的羊奶中接种的几种微生物(*Escherichia coli*, *Pseudomonas fluorescens*, *L. innocua*, *S. aureus* 和 *Lactobacillus helveticus*)的耐压性的研究表明,在压力 100～500 MPa、温度分别为 4、25、50 ℃ 的条件下处理 15 min,含脂量为 0 的羊奶中微生物对压力的抵抗性比在缓冲液中的高。但是,对于任何一种高压处理或微生物检测,奶中含脂量都没有提高微生物的抗压能力。

第 7 课 Subcritical water extraction

1. 原文

选自：Food Chemistry, 2006, 98(1): 136-148

作者：Miguel Herrero, Alejandro Cifuentes, Elena Ibañez

Subcritical water extraction (SWE), i.e. extraction using hot water under pressure, has recently emerged as a useful tool to replace the traditional extraction methods. SWE is an environmentally friendly technique[1] that can provide higher extraction yields from solid samples. SWE is carried out using hot water (from 100 to 374 ℃, the latter being the water critical temperature) under high pressure (usually from 10 to 60 bar) to maintain water in the liquid state.

The most important factor to consider in this type of extraction procedure is the variability of the dielectric constant with temperature. Water at room temperature is a very polar solvent, with a dielectric constant close to 80. However, this value can be significantly decreased to values close to 27 when water is heated up to 250 ℃, while maintaining its liquid state by applying the appropriate pressure. This dielectric constant value is similar to that of ethanol.

The experimental device[2] required for SWE is quite simple. Basically, the instrumentation consists on a water reservoir coupled to a high pressure pump to introduce the solvent into the system, an oven,[3] where the extraction cell is placed and extraction takes place, and a restrictor or valve to maintain the pressure. Extracts are collected in a vial placed at the end of the extraction system. In addition, the system can be equipped with a coolant device for rapid cooling of the resultant extract.

Although this technique has been mainly used as a batch process, studies on the on-line coupling of a SWE system to a HPLC equipment via a solid phase trapping have been reported.

Subcritical water extraction has been widely used to extract different compounds from several vegetable matrices. Likewise, one of the most deeply studied materials with SWE has been rosemary (*Rosmarinus officinalis* L.). Ibañez et al. (2003) studied the extraction of antioxidant compounds of rosemary by SWE over a wide range of temperatures. Several temperatures, from 25 to 200 ℃, were tested to study the extraction selectivity toward antioxidant compounds. There was a clear effect of water temperature on the extraction yield,[4] which increased at higher extraction temperatures. The authors verified[5] that the most polar compound (i.e. rosmanol) was the main compound extracted at low temperatures (25 ℃). When the extraction was performed at 200 ℃, a decrease in the capability of water to dissolve the most polar compounds was observed, while a high concentration of other compounds, such as carnosic acid, was obtained. Antioxidant extracts comparable to those achieved[6] using supercritical carbon dioxide extraction could be obtained by SWE. In addition to antioxidants from rosemary, the SWE extraction of aroma compounds from rosemary. Savory (*Satureja hortensis*) and peppermint (*Mentha piperita*), has also been studied。

Some studies have been conducted to compare SWE to traditional extraction methods (such as Soxhlet extraction). Clove (*Syzygium aromaticum*) extractions, performed by Clifford, Basile, and Al-Saidi (1999) demonstrated that the amount of eugenol and eugenyl acetate recovered[7] using subcritical water at 150 ℃ was similar to that achieved[8] using Soxhlet extraction and hydrodistillation. These compounds are known to possess antioxidant properties similar to those of other natural compounds, such as α-tocopherol.

In general, the use of subcritical water extraction, provides a number of advantages over traditional extraction techniques (i.e. hydrodistillation, organic solvents, solid-liquid extraction). These are, mainly, low extraction times, higher quality of the extracts (mostly for essential oils), lower costs of the extracting agent, and an environmentally compatible technique. These advantages have been verified for the SWE of several plants such as laurel, fennel, oregano and kava.

Ozel, Gogus and Lewis (2003) studied the extraction of essential oil from *Thymbra spicata*. The influences of several factors, such as temperature (100, 125, 150 and 175 ℃), pressure (20, 60 and 90 bar) and flow rate (1, 2 and 3 mL/min) were studied. It was shown that the best extraction yields (3.7%) were obtained at 150 ℃ and 60 bar, using a flow rate of 2 mL/min for 30 min. The essential oils of *Timbra spicata* were found to inhibit mycelial growth of several fungi species.

2.词或词组

clove *n*. 丁香

dielectric *n*. 电介质，绝缘体；*adj*. 非传导性的

eugenol *n*. 丁香酚

eugenyl acetate *n*. 乙酸丁香酚酯

fennel *n*. 茴香

hydrodistillation *n*. 水蒸馏法

laurel *n*. 月桂

oregano *n*. 牛至

peppermint *n*. 胡椒薄荷，薄荷油

rosemary *n*. 迷迭香

savory *n*. 香薄荷：一种地中海一年生唇形科芳香草本植物(香薄荷属，圆塔花)

soxhlet extraction 索氏提取

subcritical　*adj*.　亚临界的：小于或低于规定用数临界值的

valve　*n*.　阀，[英] 电子管，真空管

vial　*n*.　小瓶

3. 主要语法现象

[1] 定语从句

[2] 过去分词作定语

[3][4] 补语从句

[5] 宾语从句

[6][7][8] 现在分词作状语

4. 原文的参考译文

第 7 课　亚临界水萃取

在一定压力条件下用热水萃取的亚临界水萃取(SWE)方法是最近出现的用以替代传统提取方法的一种有效提取手段，它是一种能从固体物料中获得较高提取率的环境友好型的技术。SWE 是通常在 10～60 bar 的高压下，保持温度为 100～374 ℃的热水处于液体状态的条件下操作，而 374 ℃是水的临界温度。

这种萃取工艺中最重要的因素是电介常数随温度变化。在室温下，水是高极性的溶剂，介电常数接近 80。然而，当水温增加到 250 ℃时，在一定的压力下仍保持液体状态时，水的电介常数值明显的降到近 27，这个值与乙醇的电介常数相近。SWE 需要的实验装置很简单。基本部件包括与高压泵相连，向系统提供流体的水箱，用于盛装物料和进行萃取的罐，保持压力的节流阀。萃取物被收集在系统后部的小瓶中。另外，为了萃取物的快速冷却，系统中可安装制冷装置。虽然这种技术主要用于批量生产，但也有报道通过固相诱捕将 SWE 系统与 HPLC 装置结合的。

SWE 已被广泛用于一些蔬菜中不同化学成分的提取。在用 SWE 提取的材料中，迷迭香是研究最多的。Ibañez 等(2003)研究了不同温度条件下用 SWE 对迷迭香中抗氧化成分的提取，温度范围为 25～200 ℃。水温对提取量有明显的影响，高温可提高提取量。他们证实在低温(25 ℃)条件下提取的主要成分是高极性的物质，如迷迭香醇。当温度在 200 ℃ 时，发现水对极性成分的溶解能力降低了，而提取物中其他成分(如鼠尾草酸)的含量提高了。用 SWE 提取的抗氧化成分的提取量与用超临界 CO_2 提取相当。

除提取迷迭香中的抗氧化成分，SWE 还可提取迷迭香中的挥发性成分，已研究的有香薄荷、薄荷油等。一些研究将 SWE 和传统萃取方法进行了比较，对丁香的研究表明，在 150 ℃下用 SWE 提取的丁香酚和乙酸丁香酚酯的量与用索氏提取法和水蒸馏法提取的量相近。这些成分具有与某些天然成分(如维生素 E)相同的抗氧化效果。

通常，SWE 相对于传统提取技术(如水蒸馏、有机溶剂萃取、固液萃取)具有更多的优点，主要有提取时间短、萃取物品质好、费用低和环保。月桂、茴香、牛至和卡瓦等植物的 SWE 结果均证实了这些优点。

Ozel 等(2003)研究了绿薄荷(*Thymbra spicata*)挥发油的提取，研究了影响因素：温度(100、125、150、175 ℃)、压力(20、60、90 bar)和流速(1、2、3 mL/min)。结果发现在温度150 ℃、压力 60 bar、流速 2 mL/min 的条件下提取 30 min 可获得最佳的提取率(3.7%)。*Thymbra spicata* 的挥发油能抑制几种真菌菌丝体的生长。

第 8 课 Maillard reaction

1.原文

选自：Critical Reviews in Food Science and Nutrition，Food Chemistry

作者：Fabíola Cristina de Oliveira，Jane Sélia dos Reis Coimbra，Eduardo Basílio de Oliveira，Abraham Damian Giraldo Zuñiga，Edwin E. Garcia Rojas

The Maillard reaction was first described by French chemist Louis C. Maillard in 1912. The first coherent scheme of the Maillard reaction was presented by John E. Hodge in 1953.

The Maillard reaction refers to a complex group of reactions,[1] beginning with the covalent bond between the amine groups and carbonyl compounds. The Maillard reaction, for purposes of simplicity, can be divided into three stages: early, advanced and final. All these stages are interrelated and can occur simultaneously.

The early stage of the Maillard reaction is characterised by the initial glycosylation reaction. This is a condensation reaction between the carbonyl group of a reducing sugar with the available amine group,[2] which is also deprotonated, from an amino acid (or protein) to form an N-glycosylamine with the release of one water molecule. N-glycosylamine undergoes an irreversible rearrangement[3] generating the Amadori product (ARP), 1-amino-1-deoxy-ketose. With ketoses, such as fructose, a ketosylamine form, rearranges to form the product of Heyns (2-amino-2-desoxialdose). No colour changes are observed at this stage of the reaction.

The advanced stage begins with the degradation of the Amadori product, or Heyns product,[4] which may be altered by oxidation, fragmentation, enolization, dehydration, acid hydrolysis and free radical reactions,[5] resulting in multiple poorly-characterised compounds. The degradation of the Amadori product depends on system conditions, such as pH, time and temperature. The Amadori product suffers mainly 1,2 enolization with the formation of furfural (when pentoses are involved) or hydroxymethylfurfural (HMF) (when hexoses are involved) when the pH is equal to or lower than 7. At pH values above 7, the degradation of the Amadori compound involves mainly 2, 3 enolization, with reductones being formed, such as 4-hydroxy-5-methyl-2, 3-dihydrofuran-3-one, and a variety of fission products, including acetal, pyruvaldehyde and diacetyl. These compounds are highly reactive and participate in new reactions.

Carbonyl groups can be condensed with free amino groups, resulting in nitrogen incorporated in the reaction products. Dicarbonyl compounds react with amino acids, resulting in aldehydes and aminoketones. This pathway is known as Strecker degradation. Other reactions occur later in the advanced stage, including cyclisation, dehydration, retroaldolization, rearrangement, isomerization and condensation. A final phase leads to the formation of nitrogenous polymers and co-polymers of brown colouration,[6] known as melanoidins.

In contrast to the early stages, the advanced and final stages of the Maillard reaction contain a high degree of complexity. The chemistry of compounds formed in these stages are not well-known and their mechanism are not well understood,[7] although the results are easily recognized in terms of heating reactions that cause changes in the colour and flavour.

The Maillard reaction occurs naturally and may have beneficial or harmful effects on the physical, chemical, biological and organoleptic characteristics of food products in which it occurs. The harmful effects may be observed when the extent of the Maillard reaction is not controlled, e.g. during production or storage of powdered milk. However, under controlled conditions, the identity of certain products undergoing heat treatment is favoured by the Maillard reaction. This improvement is in terms of acceptance by consumers associated with the development of flavour, aroma, texture and colour in the food product,[8] which are essential for meat products, breads and grains (soybeans, peanuts, coffee and barley).

Under controlled conditions, changes in the protein structure combined with carbohydrates lead to different functionalities,[9] which are useful when the glycosylated protein is used as an ingredient to enhance functional properties such as thermal stability and solubility.

2.词或词组

Maillard reaction 美拉德反应
amine group 氨基
carbonyl group 羰基
glycosylation *n.* 糖基化
condensation *n.* 缩合
glycosylamine *n.* 葡萄糖基胺
ketosylamine *n.* 酮胺
enolization *n.* 烯醇化反应
hydroxymethylfurfural *n.* 羟甲基糠醛
melanoidin *n.* 类黑精

3.主要语法现象

[1][3][5]现在分词作状语
[2][4][8][9] 定语从句

[6] 过去分词作定语

[7] 让步状语从句

4.原文的参考译文

第8课　美拉德反应

美拉德反应由法国化学家路易斯 C.美拉德在1912年首先提出。第一个全面而完整的美拉德反应路线则由美国化学家约翰 E.霍奇于1953年提出。美拉德反应是一个复杂的反应群,起始于化合物的氨基和化合物的羰基的共价结合。为了简便起见,美拉德反应可以被分为3个阶段:早期、中期和终期。所有这些阶段是相互关联的,而且可以同时发生。

美拉德反应早期阶段以启动糖基化反应为特征。这是一个发生在还原糖的羰基与氨基酸或蛋白质的游离的、去质子化的氨基之间的缩合反应,生成N-葡萄糖基胺,同时释放一分子的水。N-葡萄糖基胺经过不可逆的重排反应,生成Amadori产物(ARP),即1-氨基-1-脱氧-2-酮糖。而果糖这类的酮糖参与,则以酮胺形式进行重排反应,生成Heyns产物,即2-氨基-2-脱氧邻酮醛糖。这个阶段的反应并没有可观察到的颜色变化。

中期阶段开始于Amadori产物或Heyns产物的降解,这个过程受氧化反应、裂解反应、烯醇化作用、脱水作用、酸水解及自由基反应的影响,会产生多种无明显特征的化合物。Amadori产物降解与反应体系的条件有关,如pH、时间和温度。当pH小于或等于7时,Amadori产物主要经1,2-烯醇化反应途径生成糠醛(五碳糖参与反应时)或羟甲基糠醛(六碳糖参与反应时)。当pH大于7时,Amadori产物主要经2,3-烯醇化反应途径,生成还原酮如4-羟基-5-甲基-2,3-二氢呋喃-3-酮,以及各种裂解产物包括乙缩醛、丙酮醛及二乙酰。这些化合物反应活性高,会参与新的反应。

羰基能与游离氨基发生缩合反应,生成含氮反应产物。二羰基化合物与氨基酸反应,生成醛和氨基酮类物质,这一反应途径称为Strecker降解。在中期末也会发生一些其他反应,包括环化、脱水、逆缩醛、重排、异构及缩合。

终期阶段形成含氮聚合物和棕褐色的共聚物,被称为类黑精。与早期阶段不同,中期和终期阶段的美拉德反应都有高度的复杂性。尽管加热反应导致颜色及风味变化的结果是如此显而易见,而这些阶段所产生的化合物却并不十分清楚,其反应机制也还没有阐明。

美拉德反应可在食品中自然发生,从而会对食品的物理、化学、生物及感官特性发挥有益的或有害的影响。在不能有效掌控美拉德反应程度时,会产生有害的影响,如在奶粉生产和贮藏过程中。在能有效掌控的情况下,可使加热过程中的美拉德反应趋于生成确定的产物。这可以改善食品产品的风味、香气、质地和颜色,随之提高消费者的喜好程度,对于肉制品、面包及谷物(大豆、花生、咖啡、大麦)是必不可少的。在有效掌控的情况下,可使蛋白质与碳水化合物结合,改变蛋白质的结构,并导致不同的功能性质。由此产生的糖基化蛋白可被用作食品成分来提高热稳定性和溶解性等功能性质。

作　业　汉译英

1. 影响淀粉改性效果的主要因素有淀粉来源、支直链淀粉的比率、颗粒形态以及用于改性的试剂的种类和浓度。淀粉特性的变化程度反映出淀粉对不同化学改性的抗拒力或敏感性。严格挑选出适宜的改性试剂和原淀粉种类可制备具有理想的特性和取代度的改性淀粉。

2. 在油脂工业中，植物油脱酸的重要性不仅在于消费者的接受度，而且对产品有极大的经济性影响。传统的脱酸工艺存在一些缺陷，一些经试验的新工艺有生物脱酸、重酯化、溶剂萃取、超临界流体萃取和膜技术。

3. 铁是一种生成血红细胞和氧化还原作用所必需的矿物元素，铁缺乏被认为是一种常见的世界性营养缺乏症。在发展中国家，降低铁缺乏症的最有效的技术措施是结合食物强化，针对高危险人群进行铁的补充，设计出对添加的和食物中本身含有的铁具有最大生物利用率的膳食策略。

4. 高压处理应用于食品加工中主要是由于它能提高食品的保藏品质。高压处理应用于食品中改善了食品各组成成分间的相互作用，影响酶反应速度，并阻止微生物的活性。高压处理可改变奶酪加工过程中乳块的凝结时间、凝乳的形成速度和奶酪产量。

5. 包装和其他与食品接触的材料中的化学物质会侵入食品中，进而被消费者摄入，这是公认的。控制这些物质是确保食品安全所必需的部分。本文综述了与包装材料相关的食品安全危害的最新知识，以及这些危害对消费者影响的评估方法。

6. 本综述讨论了用于评价高压加热杀菌和消毒对食品安全和质量方面影响的方法的现状及研究进展，研究了用高压高热的处理效果是否可达到传统的热力巴氏杀菌和消毒的处理效果，以及哪些是要特别关注的。

7. 消费者对功能食品越来越重视，使得对采用天然加工方法制备的保健成分的需求量增加。本文讨论了制备天然食品成分的环保清洁型技术，前瞻性地介绍了超临界流体萃取和亚临界水萃取这两种洁净型加工技术在用于从植物、藻类和微藻类等天然材料中分离提取天然成分的应用原理。

8. 消费者的需求促使鲜切水果和蔬菜产业不断增长。在产品加工和销售各环节均需要能保持产品质量和抑制微生物生长的新技术。单独或组合应用 C 波段紫外线、气调、热处理和臭氧处理可有效控制鲜切产品贮藏过程中微生物的生长和保持产品质量。

9. 美拉德反应起始于还原糖的羰基与氨基酸或蛋白质的氨基间的缩合反应，发展成为复杂的反应群。美拉德反应可以被分为早期、中期和终期三个阶段。终期阶段可形成含氮的、棕褐色的聚合物，被称为类黑精。美拉德反应发生在食品加工和贮藏过程中，会影响食品的物理、化学、生物及感官特性。

第 3 章

教　　材

Teaching Material

第 1 课　General processing concepts

1. 原文

选自： Principles of Food Processing(食品加工原理，影印版，中国轻工业出版社，2007)

作者： Dennis R. Heldman，Richard W. Hartel

Most food processing operations are designed to extend the shelf-life of the product by reducing or eliminating microbial activity. This general objective implies that the processing operation meets the minimum requirement of ensuring any human health safety concerns[1] associated with microbial activity. It must be acknowledged[2] that most if not all food processing operations will influence the physical and sensory characteristics of the product. It is now a common practice within the food industry to utilize processing operations as an approach to enhance the physical and sensory characteristics of food products.

Some of the general concepts associated with processing of foods include: (a) the addition of thermal energy and elevated temperatures, (b) the removal of thermal energy or reduced temperatures, (c) the removal of water or reduced moisture content, and (d) the use of packaging to maintain the desirable product characteristics established by the processing operations.

Numerous food processing operations use thermal energy to elevate product temperatures and achieve extended shelf-life. In most situations, the primary objective is the use of an established elevated temperature for some predetermined length of time to reduce the microbial population within the product. Pasteurization and blanching and commercial sterilization are excellent examples. There are several additional advantages of processes using thermal energy to elevate product temperatures. These advantages include the reduction of antinutritional components within selected products. In addition, thermal processes tend to improve the availability of several types of nutrients for human metabolism. Finally, the thermal energy provides the opportunity for good process control. It must be acknowledged that thermal processes have disadvantages as well. One of the most recognized disadvantages is the reduction in nutrient content of the product due to the thermal process.

The second processing concept to be discussed is the removal of thermal energy. The primary objective is the use of reduced temperature to reduce or eliminate microbial activity or growth during storage and distribution of the product. The chilling or reduction of product temperature followed by storage at refrigeration temperatures is used to control growth of spoilage microorganisms and achieve the desired extended shelf-life. This approach to extend shelf-life is used for many perishable products, including fresh fruits and vegetables, as well as fresh meats and seafoods. The removal of additional thermal energy leading to reduction of product temperatures below the freezing point of water leads to frozen foods and the process of food freezing. The process involves sufficient removal of thermal energy from the product to cause phase change of water within the product, inhibit microbial growth, and achieve extended shelf-life.

The third general processing concept is the removal of water from a product structure. More specifically, the process objective is the use of reduced moisture content to limit or eliminate growth of microorganisms or other factors that tend to limit product shelf-life. One of the primary categories of water-removal processes is referred to as product concentration. This process removes sufficient water from a liquid food to inhibit microbial growth. Liquid foods generally have in excess of 85% water, and concentration processes will increase the concentration of product solids to around 40% or 50%. In general, these levels of concentration limit the availability of water to microbial populations and inhibit microbial growth. The second general category of processes for removal of water includes the dehydration processes. This process provides removal of water from a food to a level where microbial activity is limited or eliminated.

The fourth concept associated with food processing is packaging or the step[3] required to maintain the product characteristics[4] established by a processing operation. Packaging materials and containers vary significantly from one product to another and are influenced by the type of processing operations used prior to packaging. The packaging or container material is selected to maintain the desirable product characteristics as established by the process.

2. 词或词组

concentration *n*. 集中,集合,专心,浓缩,浓度
dehydration *n*. 干燥
frozen food 冷冻食品
pasteurization *n*. 巴斯德杀菌法,巴氏杀菌
refrigeration *n*. 冷藏
blanching *n*. 漂烫
antinutritional component 抗营养成分
commercial sterilization 商业杀菌
spoilage microorganism 腐败菌

3. 主要语法现象

[1][3][4] 过去分词作定语,放在被修饰词的后面
[2] 主语从句

4. 原文的参考译文

第1课 基本加工概念

大多数食品加工操作是为了通过抑制或杀灭微生物的活动来延长食品的货架期而设计出

来的。此基本目的意味着食品加工操作最低限度必须抑制食品中微生物活动来保证人类的健康安全,必须承认大多数(不是所有的)食品加工操作都会影响产品的物理和感官特性。在食品工业中,利用加工操作来提高产品的物理和感官特性是普遍的做法。

与食品加工相关的一些基本概念包括:(a)增加热能和升温,(b)排除热能或降温,(c)排除水分或降低水分含量,(d)利用包装来保持加工操作所获得的良好产品特性。

许多食品加工操作采用热能来提高产品的温度,延长货架期。大多数情况下,其主要目的是利用持续一定时间的高温来降低产品中微生物的数量。巴氏杀菌、漂烫和商业灭菌都是非常好的例子。利用热能提高产品的温度还有一些优点,包括降低某些食品中抗营养成分的含量。另外,热处理会提高营养物质在人体代谢中的利用率。最后,热处理有助于很好地控制加工过程。必须承认,热处理也有缺点,其中公认的缺点之一就是热处理会造成产品营养成分的损失。

第二个加工概念是排除热能。主要目的是通过降温来抑制或消除产品贮藏和销售过程中微生物的活动或生长。冷却或降低产品温度然后再冷藏通常用来控制腐败菌的生长,延长货架期。这种延长货架期的方法通常应用于许多易腐食品,包括新鲜水果蔬菜、肉和海产品。使产品温度降到水的冰点以下可使食品冻结,生产冷冻食品。这一过程包括充分除去产品的热能使产品内的水分发生相变,抑制微生物生长,达到延长货架期的目的。

第三个加工概念是产品中水分的排除。更明确地说,加工目的是通过降低水分含量来限制或消除微生物的生长或其他影响产品货架期的因素。属于这一范畴的主要加工操作之一是浓缩,这一过程除去液态食品中的水以抑制微生物的生长。液态食品中水分含量通常超过85%,浓缩可以使其中的固形物含量达到40%或50%。一般来说,这种浓缩程度可以使食品中可被微生物利用的水分减少,抑制微生物生长。属于这一范畴的另一个主要加工操作是干制,这一过程能除去产品中的水,达到抑制或消除微生物活动的目的。

第四个加工概念是包装或用来保持加工产品特性的操作步骤。包装材料和容器依产品类型的不同有很大的差别,并受到包装前加工操作类型的影响。选择的包装或容器材料是用来保持加工过程所获得的产品的良好特性。

第2课 Membrane separations

1.原文

选自: Introduction to Food Science(食品科学导论,影印版,中国轻工业出版社,2005)

作者: Rick Parker

Membranes that have selective permeability,[1] meaning that they allow only certain molecules to pass through, have great significance for concentration in the food industry. Membranes are available[2] that can effectively separate water molecules from other food constituents,[3] resulting in liquid concentration. Other membranes are available that can separate molecules by size, resulting in both concentration and fractionation.

In each of membrane separations processes, the food material is brought into contact with one side of the membrane, and conditions adjusted so that only a portion of the material passes through the membrane. The driving force for material transport across the membrane is generally a combination of pressure forcing material across the membrane and concentration differences on either side of the membrane causing molecular diffusion of certain species.[4] By adjusting molecular type and concentration in the fluid on the opposite side of the membrane from the feed, selective separations can be obtained, as in dialysis. By far, applications of reverse osmosis and ultrafiltration dominate the use of membranes in the food industry.

Membrane processes have many advantages over other concentration techniques. The main advantage is[5] that product quality is generally maintained, since low temperatures are used and there is no vapor-liquid interface to cause loss of volatile flavors and aromas. In addition, membrane separations generally have reduced energy requirements, lower labor costs, lower floor space requirements, and wide flexibility of operation. However, membranes tend to foul as concentrated material builds up at the membrane surface and viscosity increases, and this limits concentrations that can be obtained. Generally, concentrations only between 40% to 45% can be obtained from membrane processes, as compared to over 80% for evaporation. In the past, sanitation of membranes has also been considered problem, and membrane replacement costs can be quite high. Despite these limitations, membrane separations find increasingly wider use in the food industry as new membrane materials and new membrane-based technologies are developed.

The dairy industry is one[6] that has embraced membrane separations. Reverse osmosis is used to concentrate milk and whey prior to evaporation, reduce shipping requirements for bulk transport, and produce specialty concentrates for commercial sale. Ultrafiltration is used widely to concentrate and fractionate milk prior to cheese production and to separate whey proteins from whey. Membrane processes are also employed in such diverse applications as concentrating fruit juices, wastewater streams, and other products to separating and concentrating proteins or starches from salts and sugars. Use of membrane separations in the food industry is increasing as new membrane materials and technologies are developed.

2. 词或词组

dialysis *n.* 透析,分离

fractionation *n.* 分级,分离

ultrafiltration *n.* 超滤

whey *n.* 乳清

reverse osmosis 反渗透

3. 主要语法现象

[1][3] 分词短语作同位语

[2] 定语从句

[4] 状语从句

［5］表语从句
［6］定语从句

4.原文的参考译文

第2课　膜　分　离

具有选择渗透性的膜，即它们只允许特定大小的分子通过，对食品工业的浓缩具有重大意义。膜能有效地将水同食品中其他组分分离，实现浓缩，也可依据分子大小不同将物质分开同时实现浓缩和分离。

在每次膜分离过程中，将食品物料移至膜的一侧，通过调整工艺条件只允许一部分物料通过膜。推动物料通过膜的作用力通常是强制物料通过膜的压力和膜两侧引起某种分子扩散的浓度差的合力。通过调整膜另一侧液体中分子的类型和浓度，即获得特定的分离，如透析。到目前为止，反渗透和超滤仍是食品工业中主要的膜分离方式。

与其他浓缩技术相比，膜技术有很多优点。主要的优点是普遍保持了产品的质量，因为是在低温下进行，而且没有气-液接触而引起的挥发性风味成分和香气成分的损失。此外，膜技术还具有耗能低、劳动成本低、空间需求小和操作灵活的优点。然而，浓缩物料由于在膜表面的积聚和黏性的增加，导致膜易被堵塞，这限制了浓缩的程度。通常，膜分离浓缩程度仅为40%～45%，而蒸发可达到80%。过去，膜的卫生状况也一直是值得考虑的问题，膜的更新成本非常高。尽管如此，随着新的膜材料和膜技术的发展，膜分离在食品工业的应用日益广泛。

乳品行业是采用膜分离技术的行业之一。牛乳和乳清在蒸发前用反渗透技术浓缩以减少运输容量并为商业零售生产特定的浓缩产品。超滤技术被广泛应用在干酪生产前牛乳的浓缩和分级，以及从乳清中分离乳清蛋白。膜技术还被应用在果汁浓缩、废水处理，以及其他产品如把蛋白质和淀粉从盐和糖中分离和浓缩出来。随着新的膜材料和技术的发展，膜分离技术在食品工业中的应用会更广泛。

第3课　Food-parking materials and forms

1.原文

选自：Introduction to Food Science(食品科学导论，影印版，中国轻工业出版社，2005)

作者：Rick Parker

The food industry uses four basic packaging materials: metal, plant matter (paper and wood), glass, and plastic. A number of basic packaging materials are often combined to give a suitable package.

Cans

Cans are formed at the food processing factory or shipped with their bottoms attached with separate can lids. Lids are seamed onto the cans. The outside of the steel can is protected from rust by a thin layer of tin. The inside of the can is protected by a thin layer of tin or baked-on enamel. Tin-free steel and thermoplastic adhesive-bonded seams have become more common. Aluminum is used as a packaging metal because of its light weight, low levels of corrosion (rust), recyclability, and ease of shipping. However, aluminum has less structural strength than metal cans. This limitation has been overcome by the injection of a small amount of liquid nitrogen into the can prior to closure. This gas provides for internal pressure that adds rigidity.

Glass

Glass provides a chemically inert and noncorrosive recyclable food packaging material. Glass breaks, and it is too heavy for some processing uses. Also, recycling is not easy, except in the case of home canning use.

Paper

Paper used as a primary container must be treated, coated, or laminated. Paper from wood pulp and reprocessed waste paper is bleached and coated or impregnated with waxes, resins, lacquers, plastics, and laminations of aluminum to improve its water strength and gas impermeability, flexibility, tear resistance, burst strength, wet strength, grease resistance, sealability, appearance, and printability of advertising or labels. Papers[1] treated for primary contact with food are reduced in their ability to be recycled. Paper[2] that comes in contact with foods must meet FDA standards for chemical purity. Paper used for milk cartons must come from sanitary virgin pulp. The major safety concern is with the puncturability or tearability[3] that will allow for the outside environment to enter and contaminate the food.

Plastics

Some popular plastics include cellophane, cellulose acetate, nylon, mylar, saran, and polyvinyl chloride. Copolymer plastics extend the range of useful food-packaging applications. lonomer (ionic bonds) plastic materials are improved food-handling materials that function under greater oil, grease, solvent resistance, and they have a higher melting strength. Newer plastic materials contain cornstarch, which makes them more biodegradable.

Laminates

Commercial laminates with as many as eight different layers can be custom-designed for a specific product. In the case of prepackaged dry beverages, the laminate (from outside of package to inside) may have a special cellophane that is printable, polyethylene for a moisture barrier, treated paper for stiffness, a layer for bonding, an aluminum foil (prime gas barrier) inside a layer of polyethylene for an additional water vapor barrier.

2. 词或词组

adhesive *n.* 黏合剂

cellophane *n.* 玻璃纸

copolymer *n.* 共聚物

enamel *n*. 珐琅,瓷釉; *vt*. 涂以瓷釉,彩饰珐琅,瓷釉

lacquer *n*. 漆,漆器; *vt*. 用漆涂于……,使表面光洁

lonomer *n*. 离子键聚合物

mylar *n*. 聚酯薄膜

nylon *n*. 尼龙

resin *n*. 树脂

seam *n*. 接缝,线缝,缝合线,衔接口,伤疤; *vt*. 缝合,接合,焊合,使留下伤痕

saran *n*. 莎纶,一种合成纤维

cellulose acetate 乙酸纤维素

polyvinyl chloride 聚氯乙烯

3.主要语法现象

[1] 过去分词作定语

[2][3] 定语从句

4.原文的参考译文

第3课 食品包装材料和形式

食品工业使用的4种基本的包装材料有:金属、植物原料(纸和木材)、玻璃和塑料。一个适宜的包装通常由多种基本材料共同构成。

罐

罐在食品加工厂在线成型,或成型后同盖子分开运输到食品加工厂(盖子由食品生产者在灌装后密封到罐上)。为防止生锈,铁罐的外层镀薄薄一层锡,内层镀锡或烤瓷。无锡罐和热塑黏接制罐的应用已日益普遍。铝由于具有质量轻、耐腐蚀、易回收和运输容易的特点而作为金属包装材料使用。然而,铝的强度比铁罐低,在密封前向罐中注入少量液氮可克服这一缺陷。原因是气体使罐内压升高从而增加了罐的刚性。

玻璃

玻璃是具有化学惰性、耐腐蚀、易回收的食品包装材料。但玻璃易碎、质量重不利于某些加工操作。而且,除了家庭场合使用易回收外,其他场合使用回收并不容易。

纸

纸作为初级容器必须经过处理、涂布或制成薄板。用木浆或再加工的废纸制成的纸经过漂白与涂抹,或上蜡、涂漆、涂布树脂、加入塑料和铝膜来改善其耐水强度、阻气性、柔韧性、抗撕裂性、抗爆裂性能、阻湿性、耐油性、封合性、外观和可印刷广告或标签性能。直接与食品接触的纸其重复利用的可能性有所降低。直接与食品接触的纸必须在化学纯度上达到FDA的标准。用于乳制品包装的纸必须来自卫生原纸浆。主要的安全问题与纸的抗戳性或抗撕裂性有关,戳破和撕裂会导致外部污染源进入而污染包装内的食品。

塑料

应用比较多的塑料包括玻璃纸、乙酸纤维素、尼龙、聚酯、莎纶和聚氯乙烯。聚合塑料扩大

了食品包装材料的应用范围。离子键聚合塑料是经过改善的食品用塑料材料，它具有更强的阻油性、耐溶剂性，并具有很高的熔化性能（热封性能好）。含有玉米淀粉的新塑料材料，生物降解能力更强。

复合薄膜

商业用复合薄膜最多可由8层不同材料构成，可根据用户要求为特定产品定制设计。例如，预包装固体饮料，复合薄膜构成材料（由外至内）分别是：易于印刷的玻璃纸、阻隔湿气的聚乙烯、增强硬度的纸板、粘接剂层、加强阻隔水汽性能的聚乙烯、内层的铝箔（主要用来阻隔气体）。

第4课　Description of commercial sterilization systems

1.原文

选自： Principles of Food Processing（食品加工原理，影印版，中国轻工业出版社，2007）

作者： Dennis R. Heldman, Richard W. Hartel

There are two types of heating media[1] utilized in commercial sterilization processes. The most popular is high-pressure, saturated steam at sufficiently high pressures to achieve 135 to 140 ℃. Saturated steam provides very effective heat transfer environment at the surface of the container and ensures[2] that the resistance to temperature rise within the container is dependent primarily on the container contents. Some types of containers do require careful attention when steam is brought into contact with the container surface at the beginning of the process. Careful procedures for venting the retort ensure[3] that the cavity is entirely filled with steam and without air pockets. During cooling, care is required to ensure[4] that pressure gradients within the containers are maintained at low levels to avoid damage to the containers or seals.

The second heating medium[5] used for commercial sterilization processes is hot water. Water cannot be used at temperatures above 100 ℃, but many processes for acid foods can be achieved at 100 ℃ or slightly less.

There are many different types of commercial sterilization systems. In the following, three basic categories will be described.

Batch or Still Retort. The batch or still retort is the least complex of the systems utilized for commercial sterilization of foods. The batch retort is a vessel with sufficient structural design and seal integrity to maintain the steam pressures within the vessel as required for achieving the desired process. Although the structural and seal integrities are important components of the batch or still retort, sufficient control mechanisms are required to conduct the process. The controls include the ability to adjust the steam pressure within the vessel to the desired level and to monitor the steam pressure and temperature throughout the desired process.

Continuous Retort Systems. There are a variety of different types of continuous retort systems for use in achieving commercial sterilization. Several common and unique characteristics are found in each system. Most often, the product containers are carried through the continuous system in a manner[6] so that the containers are rotating at all times. The continuous retort system may be a long cylindrical vessel containing the steam medium. The product containers enter the system through a mechanism[7] that maintains the desired pressure within the vessel. After entering the cylindrical vessel, the product containers are carried in a circular manner as they move gradually toward the exit from the cylinder. As the product containers move in circular manner, there is continuous rotation to ensure as much mixing as possible of the container contents during movement from the entrance to exit. In continuous systems, the heating medium temperature is established and maintained by the pressure of the steam within the vessel. The time for the process is established and maintained by the rate of movement of the containers from entrance to exit. In most cases, the circular pattern of movement is controlled by a central rotation mechanism, such that all containers moving through a continuous system are exposed to the heating environment for the desired period of time. The product containers leave the system through a mechanism[8] that maintains the pressure within the vessel at the required level.

Continuous Rotary Cooker. Continuous rotary cookers describe a category of systems that are operated at atmospheric pressure. These systems are used only for acid food products where process temperatures in excess of 100 ℃ are not required. Product containers are introduced into the continuous rotary cooker in a manner similar to continuous retort, and the product is carried through a cylindrical vessel with as much product container rotation as possible. The heating medium in a continuous rotary cooker may be steam or hot water spray. Both media will achieve and maintain the desired temperature levels for the duration of the thermal process. The time of the process is controlled by the rate of product container movement from entrance to exit.

2. 词或词组

batch *n*. 一批,一炉,间歇

commercial sterilization 商业杀菌

continuous retort 连续(高压)杀菌器

cooker *n*. 蒸(煮)机

cylindrical *adj*. 圆柱的

pressure gradient 气压(压力)梯度

still retort 静止式杀菌锅

3. 主要语法现象

[1][5] 过去分词作定语

[2][3][4] 宾语从句

[6] 强调句

[7][8] 定语从句

4.原文的参考译文

第4课　商业杀菌系统的类型

有2种类型的加热介质应用于商业杀菌过程中。应用最普遍的是具有足够高的压力,并可以达到135～140 ℃的高压饱和蒸汽。饱和蒸汽在产品表面创造了有效的热传递环境,容器内部温度的升高主要取决于内容物。在杀菌的起始阶段蒸汽通入,与容器表面相接触时,必须对一些类型的容器留意观察。仔细地排除杀菌锅内的冷空气,确保锅内完全充满蒸汽,没有空气残留。在冷却阶段,需要确保容器内的压力梯度维持在较低水平,以避免破坏容器或破坏容器封口。

商业杀菌过程中使用的另一种加热介质是热水。水不能用作100 ℃以上杀菌的加热介质,但酸性食品的杀菌可以是100 ℃或稍低一点。

商业杀菌系统有很多不同的类型,下面描述3种基本的类型。

间歇或静止式杀菌锅。间歇或静止式杀菌锅是最简单的食品商业杀菌系统。间歇式杀菌锅具有充分的结构设计和完整的密封性能以维持杀菌要求所需的蒸汽压力。尽管结构和密封的完整性是间歇或静止式杀菌锅的重要组成,这类杀菌还需要足够的控制装置。这些控制包括调整杀菌装置内部的蒸汽压达到所需水平,监控杀菌过程的蒸汽压力和温度。

连续蒸煮系统。有多种不同类型的连续杀菌系统可应用在商业杀菌上。各种杀菌系统既有共同点,又有自身的特点。最常见的是,装有产品的容器由连续运行的装置运载,并以一定的方式运行以使容器一直旋转。连续杀菌系统可以是一个装有蒸汽介质的长的圆柱形装置。装有产品的容器通过一个装置进入杀菌系统,容器进入时,该装置可维持杀菌器内所需的蒸汽压力。容器进入后,以环形方式被运载,从入口逐渐向出口移动。从入口移向出口的过程中,装有产品的容器以环形方式运动,连续的旋转使容器内容物尽可能混合。在连续杀菌系统中,加热介质的温度靠杀菌器内蒸汽的压力建立和维持,杀菌时间靠容器从入口移向出口的速率来确定和维持。在多数情况下,环形的运动方式由中央的旋转装置来控制,以便通过这一连续运行系统的所有容器按要求持续暴露在热环境中。装有产品的容器通过一个装置离开杀菌系统,该装置能维持杀菌锅内的压力在所需的水平。

连续旋转蒸煮机。连续旋转蒸煮机是一类在常压下操作的杀菌系统。这类杀菌系统只能用于酸性食品,杀菌温度不超过100 ℃。装有产品的容器以与连续杀菌器相似的方式被装入连续旋转蒸煮机,产品通过一个圆柱形的装置运载,同时伴随尽可能多的容器旋转。连续旋转蒸煮机内的加热介质可以是蒸汽或喷淋热水。两种加热介质在整个热处理过程中都要达到并维持所需的温度。杀菌时间由容器从入口向出口移动的速率来控制。

第 5 课　Food freezing and product quality

1.原文

选自： Principles of Food Processing（食品加工原理，影印版，中国轻工业出版社，2007）

作者： Dennis R. Heldman, Richard W. Hartel

Although the primary purpose of food freezing is[1] to serve as a preservation process and extend shelf-life, there are quality changes[2] that occur as a result of the process. In order to minimize the impact of the process, it is important to recognize the changes[3] that occur within the product during the freezing process. For example, an unfrozen product could have 70% water and 30% total solids at any temperature above the initial temperature for ice crystallization. Within a small temperature range of 5 degrees below the initial freezing temperature, a dramatic change in composition occurs. Depending on the exact composition, a product might have 30% unfrozen water, 40% ice or frozen water, and the same 30% total solids. Each degree of temperature change results in a change of product composition as the water converts to ice. As the temperature continue to decrease, the percentage of water in the frozen state as opposed to the liquid state increases. At a temperature much below the initial freezing temperature, a small fraction of the water will remain in the liquid state, and is often referred to as unfreezable water. The percentage of water[4] that is considered unfreezable is a direct function of product composition. The changes will have impact on almost all reactions occurring within the product, along with other changes[5] that have impact on product quality attributes.

Many quality attributes are influenced by the rate of freezing. If the temperature change between the initial freezing temperature and 5 degrees below the initial freezing temperature is rapid, the ice crystals[6] formed within the product structure or matrix will be small. By reducing the product temperature through this same temperature range at a slow rate, ice crystals will become much larger. The most obvious impact of freezing rate is on food texture, especially products where the water is contained within a cellular structure. In these situations, the formation of large ice crystals may disrupt cell walls and cause the loss of product structure[7] that does not recover when the product is thawed. In addition, the rupture of cell walls within product is likely to promote other reactions within the product[8] that may have negative impact on the product quality attributes.

An additional consideration in foods with a cellular structure is the diffusion of water through cell walls. Slow freezing and formation of large ice crystals encourages diffusion of water from one cell to another depending on the concentration of solutes in the unfrozen state. The movement of water from one cell to another leads to dehydration of the cell and irreversible changes due to the freezing process.

It must be recognized[9] that extremely high freezing rates may have negative impacts on product quality as well. High freezing rates[10] causing significant temperature gradients over small distances within product structure create stresses[11] that lead to stress cracking. These changes have negative impact on the product texture and should be avoided.

There are a variety of other quality attributes[12] influenced by the freezing process. In general, pigments establishing product color, flavors, and nutrients are not changed significantly due to the freezing process. On the other hand, these quality attributes may change during storage as a result of changes occurring during the freezing process. In addition, the extent of quality change during storage will be influenced dramatically by storage temperature. Other changes that occur include the destabilization of emulsions, resulting in precipitation of proteins due to the freezing process. In baked products, the amylopectin is needed to prevent starch retrogradation during the freezing process.

In summary, food product texture is the quality attribute that is influenced most dramatically by the freezing process. Other changes in quality associated with the freezing process are influenced by conditions maintained during storage of the frozen food.

2. 词或词组

amylopectin *n.* 支链淀粉
crystallization *n.* 结晶化
cellular structure 细胞结构
destabilization *n.* 扰动,不稳定
irreversible *adj.* 不可逆的
stress cracking 应力裂纹
starch retrogradation 淀粉老化
total solid 总固形物
unfrozen product 未冷冻产品

3. 主要语法现象

[1] 动词不定式作表语
[2][3][4][5][7][8][11] 定语从句
[6][10][12] 分词短语作定语
[9] 主语从句

4. 原文的参考译文

第 5 课　食品冷冻和产品质量

尽管食品冷冻的主要目的是作为一种保藏的方法来延长货架期,但是冷冻会导致产品质量发生变化。为了将冷冻对产品质量的影响减小到最低限度,了解冷冻过程中产品内部发生的变化是重要的。例如,在冻结点以上的任何温度,一个未冷冻产品可能含有70%的水分和

30%的总固形物。在冻结点以下5℃的微小温度变化区间内，产品成分发生了明显的变化。产品内部精确的组成，可能是30%的未冻结水，40%的冰或冻结水和同样是30%的总固形物。温度每改变1℃，随着水转化成冰，都会导致产品组成的变化。温度继续下降，冻结水与液体水之间的比例上升。当温度远低于冻结点时，小部分的水会以液体形式残留下来，这部分水通常被称作不能冻结的水，它所占的百分率直接受产品组成影响。这种变化连同对产品质量特性有影响的其他变化一起，对产品内部发生的几乎所有的化学反应都有影响。

许多质量特性受冻结速率的影响。如果温度从冻结点迅速下降到冻结点以下5℃，产品内形成的冰晶会很小。如果通过这个温度区间的速度慢，冰晶会变得大得多。受冻结速率影响最明显的是食品的质地，特别是水分分布在蜂窝状结构中的食品的质地。在这种情形下，大的冰晶的形成会破坏细胞壁，引起细胞结构的溃解，当产品解冻时无法再复原。另外，产品内细胞壁的破坏可能促进其他化学反应的发生，这些化学反应对产品的质量特性可能造成不良影响。

对于具有细胞结构的食品来说，另一个需要考虑的因素是水分穿过细胞壁的扩散。慢速冻结和大的冰晶的形成会促进水分在细胞间的扩散，这要取决于非冻结水中溶质的浓度。由于冷冻造成的水分在细胞间的扩散会导致细胞的脱水和不可逆的变化。

必须承认，过于快速冻结也可能对产品质量造成不良影响。冻结速度过快会造成产品内部微小距离间的温度梯度，从而产生应力，导致龟裂。这些变化对产品质地有不良的影响，应当避免。

还有许多其他的质量特性受冻结过程的影响。通常，形成产品色泽的色素、风味物质和营养成分在冷冻过程中不会发生明显改变。而在贮藏过程中，这些质量特征受冻结过程中发生的变化的影响也可能发生改变。另外，贮藏过程中质量改变的程度受贮藏温度的影响较明显。由冷冻造成的其他变化包括乳状液的不稳定，导致蛋白的沉淀。在焙烤食品生产中，需要加入支链淀粉来防止冻结过程中淀粉的老化。

综上所述，产品质地是受冷冻过程影响最明显的质量特征。其他与冷冻有关的质量变化受冻藏条件的影响。

第6课　Introduction to food additives

1.原文

选自：Food Additives (Second Edition)，(New York：Marcel Dekker，Inc.，2002)

作者：A. Larry Branen，P. Michael Davidson，Seppo Salminen，John H. Thorngate

According to the Food Protection Committee of the Food and Nutrition Board，food additives may be defined as "a substance or mixture of substances，other than a basic foodstuff，[1] which is present in a food as a result of any aspect of production，processing，storage，or packaging. The term does not include chance contaminants."

Today，more than 2 500 different additives are intentionally added to foods to produce a desired effect. Additives can be divided into six major categories：preservatives，nutritional additives，flavoring agents，coloring agents，texturizing agents，and miscellaneous additives.

There are basically three types of preservatives used in foods: antimicrobials, antioxidants, and antibrowning agents. The antimicrobials are used to check or prevent the growth of microorganisms. They play a major role in extending the shelf life of numerous snack and convenience foods. The antioxidants are used to prevent lipid and/or vitamin oxidation in food products. They are used primarily to prevent autoxidation and subsequent development of rancidity and off-flavor. They vary from natural substances such as vitamins C and E to synthetic chemicals such as butylated hydroxylanisole (BHA) and butylated hydroxytoluene (BHT). The antioxidants are especially useful in preserving dry and frozen foods for an extended period of time. Antibrowning agents are chemicals[2] used to prevent both enzymatic and nonenzymatic browning in food products, especially dried fruits or vegetables. Vitamin C, citric acid, and sodium sulfite are the most commonly used additives in this category.

Nutritional additives have increased in use in recent years as consumers have become more concerned about and interested in nutrition. These additives include vitamins, minerals, amino acids, and fiber. The vitamins are commonly added to cereals and cereal products to restore nutrients lost in processing or to enhance the overall nutritive value of the food. The addition of vitamin D to milk and of B vitamins to bread has been associated with the prevention of major nutritional deficiencies in the United States. Minerals such as iron and iodine have also been of extreme value in preventing nutritional deficiencies. The primary use of minerals is also in cereal products. Amino acids and other proteinaceous materials are not commonly used in foods. However, lysine is sometimes added to cereals to enhance protein quality. Fiber additives have been increased popularity in recent years with the increase in consumer interest in dietary fiber. Various cellulose, pectin, and starch derivatives have been used for this purpose. Recently, naturally derived fiber from apples and other fruits as well as sugarbeets has been introduced as a fiber additive.

Most coloring agents are used to improve the overall attractiveness of the food. A number of natural and synthetic additives are used to color foods. In addition, sodium nitrite is used not only as an antimicrobial, but also to fix the color of meat by interaction with meat pigments.

Flavoring agents comprise the greatest number of additives used in foods. There are three major types of flavoring additives: sweeteners, natural and synthetic flavors, and flavor enhancers. The most common additives[3] used as sweeteners are low-calorie or noncaloric sweeteners such as saccharin and aspartame. In addition to sweeteners, there are more than 1 700 natural and synthetic substances[4] used to flavor food. These additives are, in most cases, mixtures of several chemicals and are used to substitute for natural flavors. Flavor enhancers magnify or modify the flavor of foods and do not contribute any flavor on their own. Flavor enhances,[5] which include chemicals such as monosodium glutamate and various nucleotides, are often used in oriental foods or in soups to enhance the perception of other tastes.

Texturizing agents are used to add to or modify the overall texture or mouthfeel of food products. Emulsifiers and stabilizers are the primary additives in this category. Phosphates and dough conditioners are other chemicals[6] that play a major role in modifying food texture. Emulsifiers include natural substances such as lecithin and mono- and diglycerides as well as several synthetic derivatives. The primary role of these agents is to allow flavors and oils to be dispersed throughout a food product. Stabilizers include several natural gums such as carrageenan as well as natural and modified starches. These additives have been used for several years to provide the desired texture in products such as ice cream and are now also finding use in both dry and liquid products. They also are used to prevent evaporation and deterioration of volatile flavor oils.

There are numerous other chemicals used in food products for specific yet limited purposes. Included are various processing aids such as chelating agents, enzymes, and antifoaming agents; surface finishing agents; catalysts; and various solvents, lubricants, and propellants.

2.词或词组

antimicrobial　*n.*　抗微生物剂
antibrowning agent　抗褐变剂
antifoaming agent　防沫剂,消泡剂
butylated hydroxylanisole(BHA)　丁基羟基茴香醚
butylated hydroxytoluene(BHT)　二丁基羟基甲苯
carrageenan　*n.*　卡拉胶,角叉胶
catalyst　*n.*　催化剂
dough conditioner　面团改进剂
diglyceride　*n.*　甘油二酯
enzymatic browning　酶促褐变
flavoring enhancer　风味增强剂
lecithin　*n.*　卵磷脂
modified starch　改性淀粉
monoglyceride　单甘油酯
monosodium glutamate　谷氨酸钠,味精
nonenzymatic browning　非酶促褐变
nucleotide　*n.*　核苷
preservative　*n.*　防腐剂
propellant　*n.*　气雾推进剂
rancidity　*n.*　腐臭,酸败
sodium sulfite　亚硫酸钠
texturizing agent　质构化剂

3.主要语法现象

[1][5] 非限制性定语从句

[2][3][4] 过去分词作定语,放在被修饰词后面

[6] 定语从句

4.原文的参考译文

第6课　食品添加剂概述

根据食品和营养部食品安全委员会的定义,食品添加剂是指:"由于食品的生产、加工、储存或包装等过程而存在于食品中的一种物质或混合物,而不是食品基本成分。本术语不包括偶然的污染物。"

如今,有2 500多种不同种类的添加剂被有意添加到食品中以达到某种预期的效果。添加剂可分为以下6大类:防腐剂、营养增补剂、调味剂、着色剂、质构化剂和其他添加剂。

食品用的防腐剂主要有3种:抗微生物剂、抗氧化剂和抗褐变剂。抗微生物剂用来防止和抑制微生物的生长,它们在延长各种快餐食品和方便食品的货架期中起主要作用。抗氧化剂用于防止食品中的油脂和(或)维生素的氧化,其最基本的作用是阻止自动氧化和由此引发的酸败味和其他异味的产生。抗氧化剂有天然物(如维生素C和维生素E)和化学合成物[如丁基羟基茴香醚(BHA)和二丁基羟基甲苯(BHT)]之分,抗氧化剂特别适用于延长干燥食品和冷冻食品的保质期。抗褐变剂是指能够阻止食品尤其是干制果蔬制品的酶促褐变和非酶促褐变的一类化合物,维生素C、柠檬酸和亚硫酸钠是这类物质中应用最普遍的。

近年来,随着消费者对营养的关注度和兴趣的增加,营养增补剂的应用呈上升趋势。这些添加剂包括维生素、矿物质、氨基酸和膳食纤维。维生素通常添加到谷类和谷类产品中以恢复加工过程中损失的营养素或增强食品的营养。在美国,通过在牛奶中加入维生素D及在面包中添加维生素B有效防止重要营养素的缺乏。铁和碘等矿物质在防止营养缺乏方面也有重要意义。矿物质也主要应用在谷类产品中。氨基酸和其他蛋白质类物质较少应用于食品中。然而,赖氨酸有时添加到谷类食品中以提高其蛋白质质量。近年来,随着人们对膳食纤维关注程度的增加,膳食纤维应用越来越普及。各种各样的纤维素、果胶和淀粉衍生物作为膳食纤维被使用。近来,从苹果和其他水果及甜菜中所得到的天然膳食纤维也被引入膳食纤维类添加剂中。

大多数着色剂用于改善食品的整体吸引力。许多天然和化学合成类添加剂可用于食品着色。另外,亚硝酸钠不仅可作为防腐剂使用,而且可通过与肉的色素发生作用而固定肉制品的色泽。

调味剂是食品添加剂中种类最多的。有3种主要类型:甜味剂、天然和合成的风味剂及风味增强剂。应用最广泛的甜味剂是低热量或无热量的甜味剂,如糖精和阿斯巴甜。除甜味剂外,有1 700多种天然和化学合成物用于食品的调味。在大多数情况下,这些添加剂是几种化合物的混合物,而且用于替代天然风味物质。风味增强剂只能放大或修饰食品的风味,而本身不能产生任何风味。风味增强剂,包括谷氨酸单钠及各种核苷类化合物,被普遍用于东方食品

或汤中以提高其他风味。

质构化剂通常被用于增加或改善食品的整体质构或口感。乳化剂和稳定剂是这类添加剂中最主要的物质。其他化合物如磷酸盐和面团改进剂也可改善食品质地。乳化剂包括卵磷脂、单甘油酯和甘油二酯等天然物质及一些合成衍生物。它们的主要作用是使风味物质和油均匀分散于整个食品中。稳定剂包括几种天然食用胶(如卡拉胶)、天然淀粉和改性淀粉。多年来,这些添加剂被应用于冰淇淋等食品中以产生理想的质构,而现在它们也可应用于干燥和流体食品中,同时也可用于防止挥发性风味油的蒸发和腐败。

也有许多其他化合物应用于食品中,以达到某种特殊但有限的目的,这些化合物包括各种加工助剂,如螯合剂、酶制剂、防沫剂、表面加工剂、催化剂以及各种溶剂、滑润剂和气雾促进剂。

第 7 课　Major chemical processes of food deterioration

1.原文

选自: Handbook of Food Preservation(New York: Marcel Dekker, Inc., 1999)

作者: M. Shafiur Rahman

During storage and distribution, foods are exposed to a wide range of environmental conditions. Environmental factors such as pressure, temperature, humidity, oxygen, and light can trigger several reaction mechanisms[1] that may lead to food degradation. As a consequence of these mechanisms, foods may be altered to such an extent[2] that they are either rejected by or harmful to the consumer.

Several chemical changes occur during the processing and storage of foods. These changes may cause food to deteriorate by reducing its sensory and nutritional quality. Many enzymic reactions change the quality of foods. For example, fruits when cut tend to brown rapidly at room temperature due to the reaction of phenolase with cell constituents[3] released in the presence of oxygen. Enzymes such as lipoxygenase, if not denatured during the blanching process, can influence food quality even at subfreezing temperature. In addition to temperature, other environmental factors such as oxygen, water, and pH induce deleterious changes in foods[4] that are catalyzed by enzymes.

The presence of unsaturated fatty acids in foods is a prime reason for development of rancidity during storage as long as oxygen is available. While development of off-flavors is markedly noticeable in rancid foods, the generation of free radicals during the autocatalytic process leads to other undesirable reactions, for example, loss of vitamins, alteration of color, and degradation of proteins. The presence of oxygen in the immediate vicinity of food leads to increased rates of oxidation. Similarly, water plays an important role; lipid oxidation occurs at high rates at very low water activities.

Some chemical reactions are induced by light, such as a loss of vitamins and browning of meats. Nonenzymic browning is a mayor cause of quality change and degradation of the nutritional content of many foods. This type of browning reaction occurs due to the interaction between reducing sugars and amino acids,[5] resulting in the loss of protein solubility, darkening of lightly colored dried products, and development of bitter flavors. Environmental factors such as temperature, water activity, and pH have an influence on nonenzymic browning.

2. 词或词组

blanching *n*. 热烫,烫漂
catalyze *vt*. 催化,刺激,促进
degradation *n*. 腐败,降解
free radical 自由基,游离基
humidity *n*. 湿度,湿气
lipoxygenase *n*. 脂氧合酶
reducing sugar 还原糖
phenolase *n*. 酚酶
sensory quality 感官质量
trigger *v*. 引发,引起
unsaturated fatty acid 不饱和脂肪酸
water activity 水分活度

3. 主要语法现象

[1][2][4] 定语从句
[3] 过去分词作状语
[5] 独立结构,作状语

4. 原文的参考译文

第7课 食品腐败的主要化学过程

在贮藏和销售过程中,食物暴露于各种环境中。环境因素如压力、温度、湿度、氧气和光照等可以诱发一些导致食物腐败的反应。由于这些反应的发生,食物可能发生变化,被消费者拒绝,或对消费者有害。

在食物的加工和贮藏过程中常发生一些化学变化,这些变化可能会降低食品的感官品质和营养价值而使食物品质恶化。很多酶促反应也能改变食物的品质。例如,切开的水果在室温下迅速褐变。这是由于在氧气存在下,酚酶与释放出来的细胞成分发生反应。酶(如脂氧合酶)如果在食物热烫的过程中未发生变性,则它甚至在低于冰点的温度下也会影响食物的品质。除温度以外,其他环境因素如氧气、水、pH 等也可使食物在酶的催化作用下发生有害变化。

在贮藏过程中,只要氧气充足,食物中不饱和脂肪酸的存在就成为产生酸败味的首要因

素。在酸败食物中，当不良气味的产生非常明显时，在自动催化过程中产生的自由基可导致其他一些不希望发生的反应，如维生素的损失、颜色的改变和蛋白质的降解。食物中氧气的直接存在导致氧化速率的增加。类似的，水分在脂肪氧化的过程中也起着重要的作用，在极低的水分活度下脂肪迅速发生氧化。

一些化学反应是由光照引起的，如维生素的损失及肉的褐变。非酶褐变是许多食物品质变化和营养成分降解的主要原因。这一类型的褐变反应是由还原糖和氨基酸之间发生相互作用引起的，结果导致蛋白质溶解性的丧失，浅色干燥食品颜色变暗并产生苦味。温度、水分活度和 pH 等环境因素对非酶褐变都有影响。

第 8 课　The nutrients in foods

1. 原文

选自： Nutrition Concepts and Controversies (Eighth Edition) (Wadsworth: Thomson Learning Inc.)

作者： Frances Sizer, Eleanor Whitney

The body requires six kinds of nutrients—families of molecules indispensable to its functioning—water, carbohydrates, fat, protein, vitamins, and minerals. These nutrients are delivered by foods.

Foremost among the six classes of nutrients in foods is water,[1] which is constantly lost from the body and must constantly be replaced. Water acts as a solvent, provides the medium for transportation, participates in chemical reactions, provides lubrication and shock protection, and aids in temperature regulation in the human body. Among the four organic nutrients, three are energy-yielding nutrients, meaning[2] that the body can use the energy they contain. The carbohydrates and fats (fats are properly called lipids) are especially important energy-yielding nutrients. As for protein, it does double duty: it can yield energy, but it also provides materials[3] that form structures and working parts of body tissues.

The fifth and sixth classes of nutrients are the vitamins and the minerals. These provide no energy to the body. A few minerals serve as parts of body structures (calcium and phosphorus, for example, are major constituents of bone), but all vitamins and minerals act as regulators. As regulators, the vitamins and minerals assist in all body processes: digesting food; moving muscles; disposing of wastes; growing new tissues; healing wounds; obtaining energy from carbohydrate, fat, and protein; and participating in every other process necessary to maintain life.

When you eat food, then, you are not just engaging in a pleasurable activity; you are providing your body with energy and nutrients. The best food for you, then, is the kind[4] that supports the growth and maintenance of strong muscles, sound bones, healthy skin, and sufficient blood to cleanse and nourish all parts of your body. This means[5] you need food[6] that provides not only energy but also sufficient 'nutrients'. Furthermore, some of the nutrients are essential nutrients, meaning that if you do not receive them from food, you will develop deficiencies; the body cannot make these nutrients for itself. Essential nutrients are found in all six classes of nutrients. Water is essential;[7] so is a form of carbohydrate; so are some lipids, some parts of protein, all of the vitamins, and the minerals important in human nutrition too.

Nutrition profoundly affects health. Healthful nutrition can help prevent or reduce the severity of some diseases. Table 1 lists some nutrition measures to prevent diseases.

Table 1 Nutrition Measures to Prevent Diseases

Adequate Intake of Essential Nutrients,	**Moderation in Intake of Energy**
Especially Protein, and Energy from	from Food Helps Prevent
Food Helps Prevent	Obesity and related diseases, such
In Pregnancy	as diabetes and hypertension
Low birthweight	Moderation in Fat Intake Helps
Poor resistance to disease	Prevent
Some forms of birth defects	Susceptibility to obesity, some
Some forms of mental/physical	cancers, and atherosclerosis
retardation	Moderation in Sugar Intake Helps
In Infancy and Childhood	Prevent
Growth deficits	Dental caries
Poor resistance to disease	Adequate Calcium Intake Helps
In Adulthood and Old Age	Prevent
Poor resistance to infectious diseases	Adult bone loss
Susceptibility to some forms of cancer	Possibly colon cancer and
Moderation in Sodium Intake Helps	hypertension
Prevent	Adequate Iron Intake Helps Prevent
Hypertension and related diseases of the	Iron-deficiency anemia
heart and kidney	Adequate Vitamin Intake Helps
Adequate Fluoride Intake Helps	Prevent
Prevent	Susceptibility to certain cancers and
Dental caries	possibly heart disease
	Certain birth defects

2. 词或词组

anemia *n.* 贫血,贫血症

atherosclerosis *n.* 动脉硬化症

carbohydrate *n.* 糖类,碳水化合物

deficiency *n.* 缺乏,不足

diabete　*n*.　糖尿病

hypertension　*n*.　高血压

kidney　*n*.　肾

lubrication　*n*.　润滑剂

nutrient　*n*.　营养素

obesity　*n*.　肥胖

phosphorus　*n*.　磷

pregnancy　*n*.　怀孕

retardation　*n*.　延迟

susceptibility　*n*.　易感性

3. 主要语法现象

[1][3][4][6] 定语从句

[2][5] 宾语从句

[7] 倒装句

4. 原文的参考译文

第 8 课　食物中的营养素

人的身体需要 6 种营养素——水、碳水化合物、脂肪、蛋白质、维生素和矿物质，它们是人体机能中 6 种不可缺少的分子。这些营养素由食物提供。

在食物的这 6 种营养素中最重要的是水，水不断地被身体消耗，所以必须经常补充。水作为一类溶剂，是体内物质运输的介质，参与化学反应，提供润滑剂和减震保护，并且还可作为体温调节剂。在 4 种有机营养素中，3 种是可以产生能量的营养素，这意味着身体能利用它们所包含的能量。碳水化合物和脂肪（脂肪确切地说应称为脂质）是尤为重要的产能营养素。至于蛋白质，它有 2 种功能：既能产能，又能提供构建机体组织的原料。

维生素和矿物质是第 5 和第 6 种营养素。它们不能给身体提供能量。一些矿物质作为身体结构的组成成分（如钙和磷是骨骼的主要成分），但所有的维生素和矿物质都可以作为调节剂。作为调节剂，维生素和矿物质帮助调节所有的机体过程：消化食物，肌肉运动，排泄废物，生长新生组织，愈合伤口，从糖类、脂肪和蛋白质获得能量，并且参加其他一切维持生命的必需过程。

吃东西不仅仅是一件愉悦的行为，同时还在为你的身体提供能量和营养。最好的食物应该能生成并维持强健的肌肉、健全的骨骼、健康的肌肤和用以净化、滋养身体各部位的充足的血液。这意味着你所需要的食物，不仅要能提供能量，还应含有足量的营养素。而且，某些营养素还是必需的，这意味着如果你不从食物中获取它们，你将会发生相应的营养缺乏症，因为人的身体不能制造这些营养素。上述 6 种营养素均是必需营养素。水是必要的，此外，一种碳水化合物、一些脂质、部分蛋白质、全部维生素和矿物质对人体也是非常重要的。

营养对人体的健康影响很大，健康的营养能预防一些疾病或者削弱它们的危害度。表 1

列出了一些预防疾病的营养措施。

表 1　预防疾病的营养措施

必需营养素摄取足量，特别是来自于食物的蛋白质和能量 **有利于预防：** 胎儿期 出生体重过低 对疾病抵抗力弱 某些出生缺陷 某些智力或生理缺陷 幼儿和儿童期 发育不良 对疾病抵抗力弱 成人和老年期 对传染性疾病抵抗力弱 易感染某些癌症 **钠摄取适度有利于预防：** 高血压和相关的心脏与肾脏疾病 **氟摄取足量有利于预防：** 蛀牙	**能量摄取适度有利于预防：** 肥胖和相关的疾病，如糖尿病和高血压 **脂肪摄取适度有利于预防：** 肥胖症、某些癌症和动脉硬化症 **糖摄取适度有利于预防：** 蛀牙 **钙摄取足量有利于预防：** 成人骨质疏松 结肠癌和高血压 **铁摄取足量有利于预防：** 缺铁性贫血症 **维生素摄取足量有利于预防：** 某些癌症和心脏病 某些出生缺陷

第 9 课　Sensory evaluation concepts

1. 原文

选自： Sensory Evaluation of Food: Principles and Practices (Second Edition) (Springer Science + Business Media, LLC 2010)

作者： Harry T. Lawless, Hildegarde Heymann

Sensory evaluation has been defined as a scientific method[1] used to evoke, measure, analyze, and interpret those responses to products as perceived through the senses of sight, smell, touch, taste, and hearing (Stone and Sidel, 2004). The principles and practices of sensory evaluation involve each of the four activities mentioned in this definition. Consider the words "to evoke." Sensory evaluation gives guidelines for the preparation and serving of samples under controlled conditions so that biasing factors are minimized. For example, people in a sensory test are often placed in individual test booths so that the judgments[2] they give are their own and do not reflect the opinions of those around them. Samples are labeled with random numbers so that people do not form judgments based upon labels, but rather on their sensory experiences. Another example is in how products may be given in different orders to each participant to help measure and counterbalance for the sequential effects of seeing one product after another. Standard procedures may be established for sample temperature, volume, and spacing in time,[3] as needed to control unwanted variation and improve test precision.

Next, consider the words, "to measure." Sensory evaluation is a quantitative science in[4] which numerical data are collected to establish lawful and specific relationships between product characteristics and human perception. Sensory methods draw heavily from the techniques of behavioral research in observing and quantifying human responses. For example, we can assess the proportion of times[5] people are able to discriminate small product changes or the proportion of a group that expresses a preference for one product over another. Another example is having people generate numerical responses reflecting their perception of how strong a product may taste or smell. Techniques of behavioral research and experimental psychology offer guidelines as to how such measurement techniques should be employed and what their potential pitfalls and liabilities may be.

The third process in sensory evaluation is analysis. Proper analysis of the data is a critical part of sensory testing. Data generated from human observers are often highly variable. There are many sources of variation in human responses that cannot be completely controlled in a sensory test. Examples include the mood and motivation of the participants, their innate physiological sensitivity to sensory stimulation, and their past history and familiarity with similar products. While some screening may occur for these factors, they may be only partially controlled, and panels of humans are by their nature heterogeneous instruments for the generation of data. In order to assess whether the relationships observed between product characteristics and sensory responses are likely to be real, and not merely the result of uncontrolled variation in responses, the methods of statistics are used to analyze evaluation data. Hand-in-hand with using appropriate statistical analyses is the concern of using good experimental design, so that the variables of interest are investigated in a way that allows sensible conclusions to be drawn.

The fourth process in sensory evaluation is the interpretation of results. A sensory evaluation exercise is necessarily an experiment. In experiments, data and statistical information are only useful when interpreted in the context of hypotheses, background knowledge, and implications for decisions and actions to be taken. Conclusions must be drawn that are reasoned judgments based upon data, analyses, and results. Conclusions involve consideration of the method, the limitations of the experiment, and the background and contextual framework of the study. The sensory evaluation specialists become more than mere conduits for experimental results, but must contribute interpretations and suggest reasonable courses of action in light of the numbers. They should be full partners with their clients, the end-users of the test results, in guiding further research. The sensory evaluation professional is in the best situation to realize the appropriate interpretation of test results and the implications for the perception of products by the wider group of consumers to whom the results may be generalized. The sensory specialist best understands the limitations of the test procedure and what its risks and liabilities may be.

2.词或词组

sensory evaluation　感官评价
evoke　*v*.　唤起,产生,引起
counterbalance　*n*.　抗衡,平衡力；　*vt*.　抵消
sequential　*adj*.　按顺序的,相继的
perception　*n*.　感知,知觉,感觉
quantify　*v*.　确定……的数量,量化,定量
discriminate　*v*.　区别,辨别
heterogeneous　*adj*.　各种各样,成分混杂的

3.主要语法现象

[1] 过去分词作定语
[2][5] 定语从句
[3] 状语从句
[4] 宾语从句

4.原文的参考译文

第9课　感官评价概念

感官评价是一种通过视觉、嗅觉、触觉、味觉和听觉感知产品特性的科学方法,此科学方法包括4个步骤:唤起感知、测量感知、分析感知和解释感知(Stone and Sidel, 2004)。在感官评价的原理和实践的定义中提到了4种活动。首先是“唤起”。感官评价提出了应在一定的控制条件下制备和处理样品,从而使偏见因素的影响最小化。例如,感官检验员应坐在单独的有隔断的检验座位上做出独自的判断,而不受周围人员观点的影响。样品采用随机数字标记,使评价员不会依据标签而是根据他们的感官经验进行判断。还应考虑以何种顺序呈送样品给参与者,从而有助于测量并避免因顺序效应产生的影响。必须建立一个标准的操作程序,保证样品温度、品评量和间隔时间一致,以控制意外变异,提高检测的精确性。

接下来是测量。感官评价是一门定量科学,通过采集数据信息建立产品特性与人类感官之间的合理和特定关系。感官检测方法极大地借鉴了在观察和量化人类反应方面的行为研究技术。例如,我们可以估计出人们分辨产品微小变化的次数的比例或者人群中偏爱某一产品的比例。再如,人们可以对产品的味觉或嗅觉强烈程度做出数字量化反应。行为研究技术和实验心理学为如何使用测量技术及其潜在的缺点和适用范围提供了依据(准则)。

感官评价的第三步是分析。恰当的数据分析是感官检验的关键部分。人们观察得到的数据通常具有高度的可变性。造成人们反应变化的原因有很多,在感官检验中不能被完全控制。举例来说有参与者的情绪和动机、对感官刺激的生理敏感性和他们过去的经历以及对相似产品的熟悉程度。虽然针对这些因素有一些筛选方法,但只是部分被控制了,评价小组的成员仍然是根据他们天生各异的感官获得数据。为了评估观察得到的产品性质和感官反应间的联系

是真实的而不是不可控制的变量，可以采用统计学方法对评价数据进行分析。好的实验设计加上适当的统计分析，就可研究感兴趣的变量，得出合理的结论。

感官评价的第四步是结果解释。在实验中，感官评价的训练是非常必要的。只有当我们做出判断和采取措施时所进行的假设、背景知识和结论的上下关系（来龙去脉）是一致的情况下，试验中的数据和统计信息才是有用的。所得结论必须是基于数据、分析和实验结果而得到的合理判断。结论应包括所采用的方法、实验的局限性以及研究背景和前后框架。感官评价专家不仅要推导出实验结果，还必须根据数据做出解释，并提出合理的行动方案。为了指导更深入的研究，评价专家应该与检验结果的使用者——顾客成为真正的伙伴。相对于广大消费者而言，专业的感官评定者能更好地对实验结果和产品的感知做出合理解释，而广大消费者对产品的感受可能是笼统的。感官评价专家最好懂得评测过程的局限性、可能的风险以及不足之处。

作 业 汉译英

1.本文研究了漂烫、干燥方法和温度对黄秋葵某些特性的影响。与干燥样品相比，黄秋葵鲜果含有较高含量的色素，并具有较高的黏性。干燥前经过漂烫的样品与未经过漂烫的样品相比，保留了更多的色素，但是黏性降低。在真空干燥条件下，每一干燥温度下干燥的样品与采用烘箱热风干燥的样品相比，保留了更多的抗坏血酸、色素和植物黏液。干燥温度越高，对样品中色素、抗坏血酸和黏性破坏越大。

2.超滤、反渗透、渗透膜蒸馏三种膜技术联用来浓缩一种从胡萝卜中提取的天然色素——花青素的研究工作已完成。研究人员通过比较三种膜技术单独使用和联合使用所取得的色素浓缩结果，发现三种膜技术联用具有更好的浓缩效果，经过此法浓缩，色素的浓度从 40 mg/100 mL 提高到 980 mg/100 mL。

3.研究了生物降解膜延长 4 ℃条件下冷藏的鲜切莴苣的货架期的能力。试验采用 4 种膜作为被测试材料：以聚酯为基料的生物降解膜（NVT1，NVT2），铝箔和聚乙烯构成的复合膜（All-PE），聚丙烯膜（OPP）。在测试的 9 d 内对包装顶隙、微生物数量、包装莴苣的颜色进行监控。研究结果表明，被测试材料的气体渗透性是影响包装莴苣质量的主要因素。生物降解膜保证了鲜切莴苣的货架期，利用生物降解膜包装的鲜切莴苣其货架期比采用聚丙烯作为包装的处理长。

4.在这项研究工作中，模拟了静止式杀菌锅中水平放置、沿轴旋转的金属罐装黏性液体食品的杀菌过程。罐的转速为 10 r/min，设定蒸汽加热温度为 121 ℃。描绘由自然对流和强制对流产生的瞬间温度和速度变化图，并与固定（不旋转）的罐进行比较。结果表明：与先前已经在固定罐中发现的不同，自然对流和强制对流共同影响把受热最慢的区域划分为 2 个截然不同的区域。由于旋转的影响，加热结束时受热最慢的区域的体积不足旋转罐体积的 5%。

5.在这个研究中，测定了鼓风式冻结机中 1 cm^3 的花椰菜小花在不同冻结条件下的冻结时间和冻结速率。设定了 4 个温度（－20、－25、－30、－35 ℃）、6 个气流流速（70、131、189、244、280、293 m/min），测定了每个设定条件下花椰菜块的冻结速率和冻结时间。采用－20 ℃、280 m/min 冷冻的花椰菜样品的冷冻时间与采用－35 ℃、70 m/min 冷冻的花椰菜样品的冷冻时间相近。当空气流速从 70 m/min 提高到 293 m/min 时,冷冻时间大约为原来的 1/2。

6. 大多数已经存在的添加剂和所有新发现的添加剂必须经过广泛的毒理学评价以获准在食品中使用。尽管关于动物实验正确性的问题仍被提出,但是科学家一致认为动物实验确实为做出安全性决策提供了必需的信息。

7. 水果及某些蔬菜的细胞内存在酚类物质,它们在新鲜水果内是无色的,但氧化后就会变成褐色或灰色。这些酚类物质是细胞内酚酶作用的底物。当有氧气存在时,酚酶就催化细胞内的酚类物质氧化而导致变色。这种现象可通过使酶失活或排除氧气来抑制。抗氧化剂也可有效抑制酶促褐变。因为抗氧化剂比它保护的底物(酚类物质)更易氧化。

8. 营养膳食有五大特征。一是充分性:食物提供充足的各种必需的营养素、纤维素和能量。二是平衡性:所选择的食物不能过分强调一种营养素或一种食物类型而忽视了其他。三是热量控制:所选择的食物提供你需要维持适当体重所需的能量,不能多也不能少。四是适度性:所选择的食物不提供过量的脂肪、盐、糖或其他不需要的成分。五是多样性:所选择的食物每天都不同。

第 4 章

论文的题目、摘要和关键词

Title, Abstract and Keywords of Article

第1课　A new method of HACCP for the catering and food service industry

1.原文

选自：Food Control，2008，19(2)：126-134

A New Method of HACCP for the Catering and Food Service Industry

Taylor E

Abstract　This paper presents the rationale and use of a new method of[1] applying Codex HACCP principles designed specifically for caterers. It charts the process by[2] which the method was developed, set against the backdrop of international efforts to give support to initiatives[3] that more appropriately meet the needs of small and less developed businesses (SLDBs). The method was extensively piloted, evaluated and validated by the UK Food Standards Agency (FSA) and deemed compliant with 2006 EU HACCP requirements. The original 'Salford Model' was extended and published as Menu-safe, a system[4] that can be used by catering businesses of all types and sizes. Its shortened version, Safer Food Better Business (SFBB), has been developed by the FSA into a ready-to-use package for very small catering businesses.

Keywords　HACCP; Catering; Menu-safe; Safer food better business

2.词或词组

backdrop　*n*.　背景幕，(事件的)背景

caterer　*n*.　包办伙食者，备办宴会者，餐饮业者

codex　*n*.　法典，药典

compliant　*adj*.　顺从的，遵从的，适应的

deem　*v*.　认为，相信

initiative　*n*.　主动性，首创精神，行动，项目

package　*n*.　包裹，包，一揽子方案，软件包

rationale　*n*.　基本原理

ready-to-use　*adj*.　现成的，可直接使用的

Salford　索尔福德，英格兰西北部一城市

3.主要语法现象

[1] 过去分词作定语，放于被修饰名词后面

[2] 状语从句

[3] 同位语，补充说明 Menu-safe

[4] 定语从句

4.原文的参考译文

第1课 一种应用于餐饮业的HACCP新方法

摘要 本文介绍了一种新方法的基本原理及其应用。该方法应用Codex HACCP原则，专为餐饮业设计。本文用图表方式介绍了该方法的发展过程。该方法建立的背景是：国际上大力支持能更好地满足小型和欠发达餐饮业(SLDBs)需要的各项行动。该方法由英国食品标准局(FSA)进行了广泛试用、评价和审定，也符合2006年欧盟HACCP的要求。原有的"Salford模型"扩展为菜单-安全系统，它可应用于所有类型和规模的餐饮业。其简化版"食品越安全 餐饮业越好"(SFBB)，已由英国食品标准局开发成现成的方案，可在小规模的饮食业直接应用。

关键词 HACCP；餐饮业；菜单-安全；SFBB

第2课 Incidence and characterization of *Salmonella* species in street food and clinical samples

1.原文

选自： Journal of Food Safety，2007，27(4)：345-361

Incidence and Characterization of *Salmonella* Species in Street Food and Clinical Samples

Tunung R，Chai L C，Usha M R

Abstract The objectives of our study were to investigate the *Salmonella* species contamination in various types of ready-to-eat street-vended dishes or drinks，to isolate *Salmonella* spp. from clinical samples and to assess the possible relationship between the serotypes isolated from these two different environments. The isolates were characterized by their antibiotic resistance，plasmid profiles and randomly amplified polymorphic DNA (RAPD) sequences. A total of 24 salmonellae,[1] belonging to seven different serotypes, were isolated from 129 different street-vended foods and drinks and 12 clinical samples (rectal swabs). The encountered serotypes from street foods were *Salmonella biafra* ($n=8$)，*Salmonella braenderup* ($n=3$) and *Salmonella weltevreden* ($n=1$)，and from clinical samples were *Salmonella typhi* ($n=8$)，*Salmonella typhimurium* ($n=2$)，*Salmonella paratyphi* A ($n=1$) and *Salmonella paratyphi* B ($n=1$). The results showed no similarities in the types of *Salmonella* serotypes from street food and clinical samples examined. The *Salmonella* strains were resistant to one or more of the 14 tested antibiotics. Seventeen isolates harbored plasmids，with plasmid sizes ranging from 3.0 to 38.5 Mu. RAPD fingerprinting with primers OPAR3 and OPAR8 produced a combination of 21 fingerprint patterns. The dendrograms[2] generated for the *S. biafra* and *S. typhi* showed strains of the same serotypes but of very different types of sample and location of sampling clustered together,[3] indicating the possibility of cross contamination during food handling.

Keywords *Salmonella*; Antibiotic resistance; RAPD; Plasmids

2.词或词组

dendrogram *n.* 系统树图

fingerprint *n.* 指纹,手印; *vt.* 采指纹

harbor *n.* 海港,避难处; *v.* 抱有,持有

primer *n.* 引物

randomly amplified polymorphic DNA (RAPD) 随机扩增多态性 DNA

Salmonella *n.* 沙门菌属

serotype *n.* 血清型

vend *v.* 出售,贩卖

3.主要语法现象

[1][3] 现在分词短语作补语

[2] 过去分词作定语

4.原文的参考译文

第 2 课 沙门菌在街头食品和临床样品中的发生率及特征

摘要 本研究的目的是,调查街头出售的各类即食食品或饮料中沙门菌污染的情况,从临床样品中分离沙门菌,并对从这 2 种不同环境中分离的血清型之间可能存在的联系进行评价。对分离的沙门菌进行特征分析,包括其抗生素抗性、质粒组成及随机扩增多态性 DNA (RAPD)序列。从 129 种街头食品、饮料和 12 个临床样品(肛门插拭)中,共分离出 24 株沙门菌,分属于 7 个不同的血清型。从街头食品中分离的血清型有比亚法拉沙门菌($n=8$),布伦登卢普沙门菌 ($n=3$)和韦太夫雷登沙门菌($n=1$)。从临床样品中分离的血清型有伤寒沙门菌($n=8$),鼠伤寒沙门菌($n=2$),甲型副伤寒沙门菌($n=1$)和乙型副伤寒沙门菌($n=1$)。结果表明,所检验的街头食品和临床样品中沙门菌的血清型没有任何相似之处。在实验的 14 种抗生素中,这些沙门菌对一种或多种抗生素具有抗性。17 株沙门菌有质粒,质粒大小为 3.0~38.5 Mu。用 OPAR3 和 OPAR8 作引物,得到 21 种 RAPD 指纹图谱。比亚法拉沙门菌和伤寒沙门菌的系统树形图显示有血清型相同但样品类型和取样地点不同的菌株,这说明在食物处理时可能发生了交叉感染。

关键词 沙门菌;抗生素抗性;RAPD;质粒

第3课　The antimicrobial effect of thyme essential oil, nisin, and their combination against *Listeria monocytogenes* in minced beef during refrigerated storage

1.原文

选自： Food Microbiology，2008，25(1)：120-127

The Antimicrobial Effect of Thyme Essential Oil, Nisin, and Their Combination Against *Listeria monocytogenes* in Minced Beef During Refrigerated Storage

Solomakos N，Govaris A，Koidis P

Abstract　The antimicrobial effect of thyme essential oil (EO) at 0.3%，0.6%，or 0.9%，nisin at 500 or 1 000 IU/g，and their combination against *Listeria monocytogenes* was examined in both tryptic soy broth (TSB) and minced beef meat. Thyme EO at 0.3% possessed a weak antibacterial activity against the pathogen in TSB，whereas at 0.9% showed unacceptable organoleptic properties in minced meat. Thus，only the level of 0.6% of EO was further examined against the pathogen in minced meat. Treatment of minced beef meat with nisin at 500 or 1 000 IU/g showed antibacterial activity against *L. monocytogenes*，[1] which was dependent on the concentration level of nisin and the strains used. Treatment of minced beef meat with EO at 0.6% showed stronger inhibitory activity against *L. monocytogenes* than treatment with nisin at 500 or 1 000 IU/g. All treatments showed stronger inhibitory activity against the pathogens at 10 ℃ than at 4 ℃. The combined addition of EO at 0.6% and nisin at 500 or 1 000 IU/g showed a synergistic activity against the pathogen. Most efficient among treatments was the combination of EO at 0.6% with nisin at 1 000 IU/g，[2] which decreased the population of *L. monocytogenes* below the official limit of the European Union recently set at 2 log CFU/g，during storage at 4 ℃.

Keywords　Thyme essential oil；Nisin；*Listeria monocytogenes*；Minced beef

2.词或词组

essential oil　精油

nisin　*n*.　乳酸链球菌素，尼生素

Listeria monocytogenes　单核细胞增生李斯特菌

organoleptic　*adj*.　影响器官(尤指味觉、嗅觉或视觉器官)的，器官感觉的

pathogen　*n*.　致病菌，病原体

synergistic　*adj*.　增效的，协同的

thyme　*n*.　麝香草属植物，百里香

tryptic　*adj*.　胰蛋白酶的

3.主要语法现象

[1][2] 非限制性定语从句

4.原文的参考译文

第3课　百里香精油、乳酸链球菌素及其混合物对冷藏牛肉糜中单核细胞增生李斯特菌的抗菌效果

摘要　研究了0.3%、0.6%和0.9%百里香精油(EO)，500 IU/g和1 000 IU/g的乳酸链球菌素及它们的混合物对胰蛋白酶大豆肉汤(TSB)和牛肉糜中单核细胞增生李斯特菌的抑菌作用。0.3% EO对TSB中单核细胞增生李斯特菌只有微弱的抗菌活性，而当牛肉糜中加入0.9% EO时，其感官品质难以接受。因此，只在牛肉糜中加入0.6% EO，进一步研究其对李斯特菌的抑制作用。用500 IU/g或1 000 IU/g乳酸链球菌素处理的牛肉糜对单核细胞增生李斯特菌具有抑制作用，但其效果取决于乳酸链球菌素的浓度及所用的菌株。用0.6% EO处理牛肉糜的抑菌活性要优于用500 IU/g或1 000 IU/g乳酸链球菌素处理。所有处理在10 ℃时比4 ℃时的抑菌活性更好。混合加入0.6% EO和500 IU/g或1 000 IU/g乳酸链球菌素具有协同抑制作用。最有效的处理方式是将0.6% EO与1 000 IU/g乳酸链球菌素混合使用，在4 ℃贮藏，这样能够减少单核细胞增生李斯特菌的数量，至欧盟最新设定的法定限量(2 log CFU/g)之下。

关键词　百里香精油；乳酸链球菌素；单核细胞增生李斯特菌；牛肉糜

第4课　High hydrostatic pressure processing of fruit and vegetable products

1.原文

选自：Food Engineering，2005，21(4):411-425

High Hydrostatic Pressure Processing of Fruit and Vegetable Products

José A. Guerrero-Beltrán, Gustavo V. Barbosa-Cánovas, Barry G. Swanson

Abstract　High hydrostatic pressure (HHP) as a minimal thermal technology is a valuable tool for microbiologically safe and shelf-stable fruit and vegetable production. Microorganisms and deteriorative enzymes can be inhibited or inactivated[1] depending on the amount of pressure and time[2] applied to the product. The resistance of microorganisms and enzymes to pressure in fruit and vegetable products also is dependent on both the type

and the amount of enzymes or microorganisms as well as food composition. While on one hand, microorganisms (other than spores) can be inactivated at mild pressures (<300 MPa), on the other, enzymes can be very resistant to pressure and their resistance may increase when isolated forms are pressurized. Nevertheless, microbiologically safe fruit and vegetable products can be obtained without change in flavor if temperature is not increased beyond pasteurization temperatures. The remaining enzyme activity in HHP processed fruit and vegetable products can be delayed if a combination of obstacles, such as refrigeration temperatures, low pH, and antibrowning agents, are used to increase the shelf life of these types of products. Therefore, HHP is a promising minimal thermal technology[3] that can be used to deliver more variety of less processed fruit and vegetable products than consumers are demanding today.

Keywords　High pressure; Fruit and vegetables; HHP and enzymes; HHP and microorganisms

2. 词或词组

antibrown　*v.*　抗褐变

deteriorative　*adj.*　变质性的，恶化的

hydrostatic　*adj.*　静水力学的，流体静力学的

pressure　n. 压力，压迫感；*vt.*　迫使，增压

pasteurization　*n.*　巴氏杀菌法

3. 主要语法现象

[1] 现在分词作状语

[2] 过去分词作定语

[3] 定语从句

4. 原文的参考译文

第 4 课　果蔬产品的高静水压加工

摘要　高静水压(HHP)是一种最低热处理技术，是保证果蔬微生物安全和货架期稳定的重要手段。对产品施加一定的压力并保持一定的时间，微生物和引起变质的酶会受到抑制或失活。果蔬中微生物和酶对压力的抗性也取决于酶或微生物的类型、数量，以及果蔬的组成成分。一方面，微生物(孢子除外)在较小的压力(<300 MPa)下即可失活；另一方面，酶可能会具有很强的抗压能力，而且当单独对酶进行加压时，抗性会增加。不过，当温度不超出杀菌温度时，可得到风味无任何损失而且微生物安全的果蔬产品。如果综合应用低温、低 pH、抗褐变试剂等手段，可以增加产品的货架期，经 HHP 处理后的果蔬产品的酶活会进一步降低。因此，HHP 是一种很有发展前途的最低热处理技术，可为消费者提供更多种仅需少量加工的果蔬产品。

关键词　高压；果蔬；HHP 和酶；HHP 和微生物

第5课 Osmotic dehydration of pineapple as a pre-treatment for further drying

1.原文

选自：Journal of Food Engineering，2008,85(2):277-284

Osmotic Dehydration of Pineapple As a Pre-treatment for Further Drying

Lombard G E, Oliveira J C, Fito P

Abstract Osmotic dehydration can be used as a pre-treatment for tropical fruits with the final aim of obtaining high quality[1] dried fruit products. This concept was applied to South African grown Cayenne type pineapple pieces. The effect of osmotic dehydration on mass fluxes (water loss, solids gain and weight reduction) was investigated. Pineapple cylinders of 2 cm in diameter and 1 cm thick were immersed in sucrose solutions of 45,55 and 65 °Brix at 30,40 and 50 ℃ for 20,40,60,20,180 and 240 min. Experiments were conducted at both atmospheric pressure and applying a 200 mbar vacuum pulse during the first 10 min. Water loss and solids gain increased with temperature and concentration.[2] Applying a vacuum pulse facilitated water loss especially at the highest concentration and temperature. Furthermore, the yield was improved by applying a vacuum pulse, as mass loss was less in those cases. Temperature affected mostly the water loss while the concentration of the solution affected mostly the solids gain. Prototypes of high/low water loss and high/low solids gain combinations were selected for quality evaluation.

Keywords Pineapple; Osmotic dehydration; Vacuum impregnation; Translucency; Quality

2.词或词组

cayenne type pineapple 卡因类菠萝

osmotic dehydration 渗透脱水

pineapple cylinder 菠萝圆片

prototype *n.* 原型

3.主要语法现象

[1] 过去分词作定语

[2] 现在分词作主语

4.原文的参考译文

第5课 菠萝渗透脱水作为进一步干燥的预处理

摘要 渗透脱水可用于热带水果的预处理，其最终目的是获得高品质的水果干制品。我们将这一概念应用于南非种植的卡因菠萝片上，研究了渗透脱水对质量流量（脱水、固形物增加、以及重量减少）的影响。将直径2 cm，厚1 cm的菠萝圆片浸入在温度为30、40、50 ℃，浓度为45、55和65°Brix的蔗糖溶液中20、40、60、80、120、180和240 min。前10 min分别采用常压和200 mbar的真空脉冲的方式进行试验。随着温度和浓度的增加，脱水和固形物含量均增加。真空脉冲加快了脱水，特别是在高温和高浓度时。而且，在这些情况下质量损失较少，因此应用真空脉冲使得产量增加。温度主要影响脱水，而溶液浓度主要影响固形物含量。高/低水分损失和高/低固形物含量结合的技术原型已用于品质评价。

关键词 菠萝；渗透脱水；真空浸渍；半透明；品质

第6课 Free-radical-scavenging activity and total phenols of noni (*Morinda citrifolia* L.) juice and powder in processing and storage

1.原文

选自：Food Chemistry，2007，102(1)：302-308

Free-radical-scavenging Activity and Total Phenols of Noni (*Morinda citrifolia* L.) Juice and Powder in Processing and Storage

Yang J, Paulino R, Janke-Stedronsky S

Abstract The fresh juice of noni (*Morinda citrifolia* L.),[1] a tropical plant[2] used as a folk medicine in Pacific islands, possessed free-radical-scavenging activity (RSA), 1,1-diphenyl-2-picrylhydrazyl (DPPH), at 140 mg equivalent ascorbic acid/100 mL and total phenols at 210 mg gallic acid/100 mL. Fermentation of noni fruit for 3 months resulted in a loss of more than 90% of RSA. Dehydration at 50 ℃ produced a loss of 20% of RSA. Storage of fresh noni juice at 24 ℃ for 3 months reduced RSA more than 90%. Storage of noni juice or powder at 18 ℃ and 4 ℃ for 3 months decreased RSA by 10%-55%. The reduction of RSA of noni juice or powder during heat treatment or dehydration was much greater than reduction of total phenols. For maintenance of the substantial antioxidant properties of noni products, processing of noni powder or fresh frozen noni juice rather than fermented noni juice is recommended.

Keywords Noni juice; Noni powder; Radical-scavenging activity; Antioxidants; Total phenols

2.词或词组

antioxidant *n*. 抗氧化剂; *adj*. 抗氧的

1,1-diphenyl-2-picrylhydrazyl 二苯代苦味酰基

free-radical-scavenging 自由基清除

gallic acid 五倍子酸,没食子酸

noni *n*. 诺丽(果),海巴戟,橘叶巴戟

3.主要语法现象

[1] 同位语

[2] 过去分词作定语

4.原文的参考译文

第6课 诺丽(*Morinda citrifolia* L.)果汁和果粉的自由基清除能力和总酚含量在加工和贮藏中的变化

摘要 诺丽(*Morinda citrifolia* L.)是太平洋诸岛上一种用作民间药物的热带植物,其鲜汁具有清除自由基的能力(RSA),其清除二苯代苦味酰基(DPPH)自由基的能力相当于140 mg/100 mL抗坏血酸,其总酚含量为210 mg五倍子酸/100 mL。诺丽果经3个月发酵后RSA损失90%以上,50 ℃干燥后RSA损失20%。鲜果汁在24 ℃贮藏3个月,RSA损失90%以上。诺丽果汁和果粉在−18 ℃和4 ℃贮藏3个月,RSA下降10%~55%。在热处理或干燥过程中,诺丽果汁或果粉中RSA含量的下降比总酚含量下降大得多。为保持诺丽产品的抗氧化能力,建议加工成诺丽果粉和新鲜冷冻果汁而非发酵果汁。

关键词 诺丽果汁;诺丽果粉;自由基清除能力;抗氧化剂;总酚

第7课 Synthesis and characterization of canola oil-stearic acid-based trans-free structured lipids for possible margarine application

1.原文

选自:Journal of Agricultural and Food Chemistry,2007,55(26):77-89

Synthesis and Characterization of Canola Oil-Stearic Acid-Based Trans-Free Structured Lipids for Possible Margarine Application

Lumor S E, Jones K C, Ashby R

Abstract　Incorporation of stearic acid into canola oil to produce trans-free structured lipid (SL) as a healthy alternative to partially hydrogenated fats for margarine formulation was investigated. Response surface methodology was used to study the effects of lipozyme RM IM from *Rhizomucor miehei* and *Candida rugosa* lipase isoform 1 (LIP1) and two acyl donors, stearic acid and ethyl stearate, on the incorporation. Lipozyme RM IM and ethyl stearate gave the best result. Gram quantities of SLs were synthesized[1] using lipozyme RM IM, and the products were compared to SL[2] made by chemical catalysis and fat from commercial margarines. After short-path distillation, the products were characterized by GC and RPHPLC-MS to obtain fatty acid and triacylglycerol profiles, ^{13}C NMR spectrometry for regiospecific analysis, X-ray diffraction for crystal forms, and DSC for melting profile. Stearic acid was incorporated into canola oil, mainly at the *sn*-1, 3 positions, for the lipase reaction, and no new trans fatty acids formed. Most SL products did not have adequate solid fat content or β'-crystal forms for tub margarine, although these may be suitable for light margarine formulation.

Keywords　Canola oil; Interesterification; Lipozyme RM IM; Response surface methodology; Sodium methoxide; Stearic acid

2. 词或词组

canola　*n*.　经基因改良的油菜(油中芥子酸含量很低)

diffraction　*n*.　衍射

margarine　*n*.　人造奶油

regiospecific　*adj*.　位置特异性的,位置选择性的

short-path distillation　短程蒸馏

stearic acid　硬脂酸,十八(烷)酸

triacylglycerol　*n*.　甘油三酯

3. 主要语法现象

[1] 现在分词作状语

[2] 过去分词作定语

4. 原文的参考译文

第7课　可能应用于人造奶油的菜籽油硬脂酸基无反式脂肪酸的结构油脂的合成及其特性

摘要　无反式脂肪酸的结构油脂(SL)可作为保健成分,替代人造奶油配方中部分氢化的油脂。本文研究了用硬脂酸和菜籽油反应制取这种结构油脂。用响应面法研究了来自米黑根毛霉的脂肪酶RM IM和皱褶假丝酵母亚型1脂肪酶(LIP1)、2个酰基供体、硬脂酸和硬脂酸乙酯对反应的影响。脂肪酶RM IM和硬脂酸乙酯的效果最好。用脂肪酶RM IM合成了克级的结构油脂,该产品可与化学催化法制取的结构油脂和商业人造奶油中的脂肪媲美。该产品经短程蒸馏后,用气相色谱和反相高效液相色谱-质谱测定产物中脂肪酸和甘油三酯的组成,用^{13}C-NMR进行位置选择性分析,用X射线衍射分析晶体形态,用差示量热扫描仪测定其熔化模式。在有脂肪酶时,硬脂酸与菜籽油的反应主要在*sn*-1,3位置进行,不会形成新的反式脂肪酸。无反式脂肪酸的结构油脂产品可能适于配制软质人造奶油(light margarine),但是大多没有足够的固体脂肪含量或β'-晶体形态,不适于配制桶装人造奶油(tub margarine)。

关键词　菜籽油;酯交换;脂肪酶RM IM;响应面法;甲醇钠;硬脂酸

第8课　Fractionating soybean storage proteins using Ca^{2+} and $NaHSO_3$

1. 原文

选自:Journal of Food Science, 2006, 71(7):413-424

Fractionating Soybean Storage Proteins Using Ca^{2+} and $NaHSO_3$

Dea N A, Murpe P A, Johson L A

Abstract　Individual soybean storage proteins have been identified as having nutraceutical properties, especially β-conglycinin. Several methods to fractionate soy proteins on industrial scales have been published, but there are no commercial products of fractionated soy proteins. The present study addresses this problem by using calcium salts to achieve glycinin-rich and β-conglycinin-rich fractions in high yields and purities. A well-known 3-step fractionation procedure[1] that uses SO_2, NaCl, and pH adjustments was evaluated with $CaCl_2$ as a substitute for NaCl. Calcium was effective in precipitating residual glycinin, after[2] precipitating a glycinin-rich fraction, into an intermediate fraction at 5 to 10 mmol/L $CaCl_2$ and pH 6.4, eliminating the contaminant glycinin from the β-conglycinin-rich fraction. Purities of 100% β-conglycinin with unique subunit compositions were obtained after prior precipitation of the glycinin-rich and intermediate fractions. The use of 5 mmol/L SO_2 in combination with 5 mmol/L $CaCl_2$ in a 2-step fractionation procedure produced the highest purities in the glycinin-rich (85.2%) and β-conglycinin-rich (80.9%) fractions. The glycinin in the glycinin-rich fraction had a unique acidic (62.6%) to basic (37.4%) subunit distribution. The β-conglycinin-rich fraction was approximately evenly distributed among the β-conglycinin subunits (30.9%, 35.8%, and 33.3%, for α′-, α′-, and β-subunits, respectively). Solids yields and protein yields, as well as purities and subunit compositions, were highly affected by pH and SO_2 and $CaCl_2$ concentrations.

Keywords Soybean storage proteins; Glycinin; β-conglycinin; Fractionating

2. 词或词组

β-conglycinin *n.* β-大豆伴球蛋白
controlled atmosphere storage 气调贮藏
fractionate *vt.* 分组,分馏,分部分离
glycinin *n.* 大豆球蛋白
mangaba *n.* 芒格巴(果),一种藤本植物
nectarine *n.* 油桃
nutraceutical *adj.* 营养保健品(的)
pectinlyase *n.* 果胶裂解酶
pectin methylesterase *n.* 果胶甲酯酶
polygalacturonase *n.* 多聚半乳糖醛酸酶
precipitate *n.* 沉淀物; *vt.* 使沉淀

3. 主要语法现象

[1] 定语从句
[2] 现在分词作介词宾语

4. 原文的参考译文

第8课 用 Ca^{2+} 和 $NaHSO_3$ 分步分离大豆贮藏蛋白

摘要 现已鉴定个别大豆贮藏蛋白,特别是β-大豆伴球蛋白,具有营养保健性质。虽然已经报道了几种工业化分离大豆蛋白的方法,但是并没有商业化的大豆分离蛋白产品出现。本研究针对这个问题,用钙盐处理获得了富含大豆球蛋白和β-大豆伴球蛋白的组分,其得率和纯度都较高。用氯化钙替代氯化钠,对人们熟知的三步分离法(使用二氧化硫、氯化钠,调整pH)进行了评价。将富含大豆球蛋白的组分沉淀后,在5~10 mmol/L $CaCl_2$、pH 6.4条件下,钙能将剩余的大豆球蛋白有效地沉淀形成中间组分,这样可将富含β-大豆伴球蛋白部分中的大豆球蛋白杂质除去。将富含大豆球蛋白的组分和中间组分先行沉淀后,即可得到纯度为100%的具有独特亚基组成的β-大豆伴球蛋白。在二步分离法中,结合使用5 mmol/L SO_2和

5 mmol/L $CaCl_2$可得到最高纯度的富含大豆球蛋白(85.2%)和β-大豆伴球蛋白(80.9%)的组分。在富含大豆球蛋白的组分中,大豆球蛋白分布有特殊的酸性(62.6%)和碱性(37.4%)亚基。在β-大豆伴球蛋白组分中,亚基分布大体均衡(α′-,α′-,β-亚基分别为30.9%,35.8%,33.3%)。固形物和蛋白质的得率及其纯度和亚基组成,受pH、SO_2和$CaCl_2$浓度的影响很大。

关键词 大豆贮藏蛋白;大豆球蛋白;β-大豆伴球蛋白;分部分离

第9课 Quality changes in fresh-cut peach and nectarine slices as affected by cultivar, storage atmosphere and chemical treatments

1.原文

选自: Journal of Food Science, 1999, 64(3): 429-432

Quality Changes in Fresh-cut Peach and Nectarine Slices as Affected by Cultivar, Storage Atmosphere and Chemical Treatments

Gorny J R, Hess-Pierce B, Kader A A

Abstract The shelf-life of slices from 13 cultivars of peaches and 8 cultivars of nectarines, varied (between 2 and 12 days at 0 ℃). Controlled atmospheres of 0.25 kPa O_2 and/or 10 kPa or 20 kPa CO_2 extended the shelf-life at 10 ℃ of 'O'Henry' or 'Elegant Lady' peach slices by 1-2 days beyond the air control. Low (0.25 kPa) O_2 acted synergistically with CO_2 levels of 10 or 20 kPa to induce fermentative metabolism as indicated by ethanol and acetaldehyde production. A 2% (*W/V*) ascorbic acid 1% (*W/V*) calcium lactate postcutting dip resulted in limited reduction of cut surface browning and tissue softening in 'Carnival' peach slices.

Keywords Controlled atmosphere; Fresh cut; Stone fruit; Peaches; Nectarines

2.词或词组

controlled atmosphere storage 气调贮藏

nectarine *n*. 油桃

3.原文的参考译文

第9课 品种、贮藏气体和化学处理对鲜切桃片和油桃片品质的影响

摘要 13个品种的桃子和8个品种的油桃的切片在0 ℃贮藏时,货架期各不相同(2~12 d)。在10 ℃时,采用0.25 kPa O_2和(或)10 kPa或20 kPa CO_2的气调贮藏,可以将

O'Henry 或 Elegant Lady 桃片的货架期延长 1～2 d。低 O_2(0.25 kPa)与 10 kPa 或 20 kPa CO_2 可协同诱导发酵型代谢，产生乙醇和乙醛。Carnival 桃切片后，用 2%(*m/V*) 抗坏血酸和 1%(*m/V*)乳酸钙溶液浸泡，表面褐变和组织软化稍有减少。

关键词　气调贮藏；鲜切；核果；桃；油桃

第 10 课　Pectinolytic enzymes secreted by yeasts from tropical fruits

1. 原文

选自： FEMS Yeast Research, 2005, 5(9): 859-865

Pectinolytic Enzymes Secreted by Yeasts From Tropical Fruits

Evânia Geralda da Silva, Maria de Fátima Borges, Clara Medina

Abstract　Three hundred yeasts[1] isolated from tropical fruits were screened in relation to secretion of pectinases. Twenty-one isolates were able to produce polygalacturonase and among them seven isolates could secrete pectinlyase. None of the isolates was able to secrete pectin methylesterase. The pectinolytic yeasts identified belonged to six different genera. *Kluyveromyces wickerhamii* isolated from the fruit mangaba (*Hancornia speciosa*) secreted the highest amount of polygalacturonase,[2] followed by *K. marxianus* and *Stephanoascus smithiae*. The yeast *Debaryomyces hansenii* produced the greatest decrease in viscosity while only 3% of the glycosidic linkages were hydrolysed,[3] indicating[4] that the enzyme secreted was an *endo*-polygalacturonase. The hydrolysis of pectin by polygalacturonase secreted by *S. smithiae* suggested an *exo*-splitting mechanism. The other yeast species studied showed low polygalacturonase activity.

Keywords　Pectinases; Polygalacturonases; Pectinlyases; Tropical fruits

2. 词或词组

polygalacturonase　*n.*　多聚半乳糖醛酸酶

pectinlyase　*n.*　果胶裂解酶

pectin methylesterase　*n.*　果胶甲酯酶

3. 主要语法现象

[1] 现在分词作定语

[2] 过去分词作非限制性定语

[3] 现在分词作补语

[4] 宾语从句

4. 原文的参考译文

第 10 课　热带水果酵母分泌的果胶酶

摘要　从分离自热带水果的 300 个酵母菌中筛选分泌果胶酶的菌株。21 株能产生聚半乳糖醛酸酶，其中 7 株能分泌果胶裂解酶。没有菌株能分泌果胶甲酯酶。已鉴定的产果胶酶的酵母菌属于 6 个不同的属。从芒格巴果（*Hancornia speciosa*）中分离的威克汉姆克鲁维酵母分泌聚半乳糖醛酸酶的量最高，其次是马克斯克鲁维酵母和史密斯冠孢酵母。汉逊德巴利酵母在只有 3% 糖苷键水解时黏度降低最多，表明其分泌的酶是内切聚半乳糖醛酸酶。史密斯冠孢酵母分泌的聚半乳糖醛酸酶是以外切方式水解果胶。其他酵母多聚半乳糖醛酸酶的活性较低。

关键词　果胶酶；聚半乳糖醛酸酶；果胶裂解酶；热带水果

作　业　汉译英

1. 以草莓为试材，用 50 kV/m、100 kV/m 高压静电场处理草莓果实 30 min 后，在 4～8 ℃ 低温，RH 为 90% 左右条件下贮藏 12 d，研究高压静电场处理对其生理生化指标的影响。结果表明，经高压静电场处理的草莓果实腐烂指数、丙二醛含量、多酚氧化酶活性、花青素含量均明显低于对照，其中以 100 kV/m 处理 30 min 的效果最佳。另外，高压静电场不利于保持草莓果实外观品质。

2. 采用 Alcalase 和 Neutrase 双酶分步水解法，制备了水解度约为 24% 的大豆多肽，用截留分子质量分别为 30 000、10 000、5 000 u 的超滤膜将其分离成 4 个分子质量段，对不同分子质量段大豆多肽的溶解性、起泡性及起泡稳定性、乳化性及乳化稳定性、抗氧化性和 ACE 抑制活性进行了研究。

3. 随着转基因食品的推广应用，人们越来越关心其食用安全性。以转基因玉米及其亲本为实验材料，借助近红外光谱仪对转基因玉米及其亲本进行了识别分析：扫描区间为 12 000～4 000 cm^{-1}，分辨率为 4 cm^{-1}，扫描次数为 64 次。近红外分析还具有无污染、成本低等优点，是一种极具前景的转基因食品安全检测识别技术。

4. 讨论了面包酵母耐冷冻能力的评价方法，通过对酵母在普通面团中冷冻后的相对发酵力、在液体模拟面团中冷冻后的相对发酵力和酵母细胞冷冻存活率之间的比较，发现三者之间具有显著的线性正相关性。而应用液体模拟面团测定酵母耐冷冻能力的方法简便、易控制，可用于耐冷冻酵母菌种的选育和耐冷冻酵母产品的检测。

5. 通过比较正常浓缩苹果汁和二次混浊浓缩汁的主要化学成分，揭示浓缩苹果汁二次混浊的主要成分。结果表明：二次混浊浓缩汁总酚、缩合单宁含量高于正常浓缩汁，而小分子酚类含量低于正常浓缩汁。二次混浊浓缩汁可溶性蛋白质含量和金属离子 K^+、Ca^{2+}、Mg^{2+} 含量也高于正常浓缩汁。

6. 采用微波对 30% 水分含量的马铃薯淀粉进行处理，研究了微波辐射前后马铃薯淀粉物化性质的变化。结果表明微波淀粉颗粒表面出现凹坑，降低了膨胀度和溶解度、冻融稳定性。主要的 X 射线衍射峰的强度增大，晶型由 B 型转为 A 型，马铃薯淀粉经处理后糊化起始温度升高、黏度降低，黏度曲线由 A 型变为 C 型。微波处理使淀粉分子发生一定程度的降解。

7. 固相微萃取 (SPME)是在固相萃取 (SPE)的基础上发展起来的新型萃取分离技术。该技术集采样、萃取、浓缩、进样于一体，简便、快速、经济安全、无溶剂、选择性好且灵敏度高。文中介绍了固相微萃取的装置、原理、操作，论述了其工作条件选择和优化，以及在食品分析中的应用，并对其前景进行了展望。

8. 嗜热链球菌和保加利亚乳杆菌是酸乳生产中常用的发酵菌种，二者由于存在共生关系，较难分离。文中采用改进的 Elliker 培养基和酸化 MRS 培养基，将试管法与平板法作比较，并将试管法应用于酸乳发酵过程中乳酸菌的检测。研究表明，使用试管法代替目前常用的平板倾倒法，能够很好分离 2 种共生菌；试管法比起平板法具有操作简便、结果准确、染杂菌风险小等优点。该法测得结果的重复性好，而且菌落总数检出率较平板法高，是一种高效简便的乳酸菌活菌计数方法。本法也可广泛应用于其他兼性厌氧菌的分离检测工作。

第 5 章

论文的前言

Introduction of Article

第 1 课 Reduced and high molecular weight barley beta-glucans decrease plasma total and non-HDL-cholesterol in hypercholesterolemic Syrian golden hamsters

1.原文

选自： Journal of Nutrition,2004,134(10):2617-22

作者： Wilson T A, Nicolosi R J, Delaney B, Chadwell K, Moolchandani V, Kotyla T, Ponduru S, Zheng G H, Hess R, Knutson N, Curry L, Kolberg L, Goulson M, Ostergren K

Clinical trials have demonstrated that consumption of oats or barley lowers serum cholesterol concentrations. Similar results have been reported in hamsters fed hypercholesterolemic diets. The substance[1] present in the soluble fiber fraction of both cereal grains[2] to which this effect has been attributed is (1,3)(1,4)-β-*D*-glucan (β-glucan). Clinical and animal studies[3] that[4] used concentrated β-glucan preparations from oats and barley showed similar properties in humans and hypercholesterolemic hamsters. Structural differences between oat and barley β-glucan have been reported,[5] as have differences between molecular weight (MW) and solubility, but the cholesterol-lowering properties are approximately equivalent.

A distinguishing feature of native β-glucan from either oats or barley is the high MW ($\geqslant$1 000 ku). Because the high MW of β-glucan contributes to high viscosity in food applications and undesirable sensory properties, production of reduced MW β-glucan preparations has been considered. It was shown[6] that preparations of β-glucan concentrates from barley cultivars with low viscosity do not possess cholesterol-lowering activity. In addition, reduced MW β-glucan concentrates[7] that are likely to possess more desirable sensory properties have been prepared either by addition of exogenous β-glucanase enzymes or by manipulation of endogenous β-glucanase activity during production. However, these studies reported that enzymatic hydrolysis of high MW β-glucan either reduces or eliminates the cholesterol-lowering activity. The relation between the MW of β-glucan and its cholesterol-lowering activity is difficult to establish from these previous studies because the MW was not determined either before or after enzymatic treatment. Recent studies suggest that β-glucan concentrates with intermediate MW may, in fact, possess cholesterol-lowering activities similar to those of high MW β-glucan. Therefore, while the evidence suggests[8] that consumption of high MW β-glucan can lower plasma cholesterol concentrations and[9] that consumption of reduced MW β-glucan does not, it cannot be determined from the available studies[10] whether there is any cholesterol-lowering activity associated with consumption of a β-glucan concentrate with a MW between that found in the grain and the enzymatically degraded β-glucan.

The present study was conducted to evaluate the possibility that the hypocholesterolemic activity of β-glucan exists at both high and low MW. Concentrated high MW β-glucan (MW = 1 000 ku) was prepared from barley. A reduced MW β-glucan concentrate (MW = 175 ku) was produced from the same barley flour by enzymatic digestion with β-glucanase. The effect of dietary consumption of these preparations on lipoprotein cholesterol, plasma lipids, fecal excretion of neutral sterols, and aortic cholesterol accumulation in hamsters fed a hypercholesterolemic diet (HCD) was evaluated.

2. 词或词组

accumulation *n.* 积聚,堆积物
aortic *adj.* 大动脉的
clinical trial 临床试验
cholesterol-lowering activity 降低胆固醇效果
cultivar *n.* 栽培变种,培育植物
glucan *n.* 葡聚糖
glucanase *n.* 葡聚糖酶
hypercholesterolemic *adj.* 高胆固醇的
intermediate *adj.* 中间的; *n.* 媒介
lipoprotein *n.* 脂蛋白
neutral sterol 中性固醇
viscosity *n.* 黏质,黏性

3. 主要语法现象

[1][4] 分词短语作定语
[2][7] 定语从句
[3][8][9] 宾语从句
[5] 同位语
[6][10] 主语从句

4. 原文的参考译文

第 1 课 低分子量和高分子量的大麦 β-葡聚糖降低患高胆固醇血症的叙利亚金鼠血中总胆固醇和非高密度脂蛋白胆固醇浓度

临床研究表明,食用燕麦或大麦能降低血清胆固醇浓度,在喂食高胆固醇饲料的大鼠实验中发现相似效果。产生此生理作用的物质是这 2 种谷物的可溶纤维中都含有的(1,3)-(1,4)-β-*D*-葡聚糖(β-葡聚糖)。临床和动物实验都证明,从燕麦和大麦中分离出的β-葡聚糖对人和高胆固醇大鼠具有相同效果。有报道燕麦和大麦的β-葡聚糖存在结构差异,正如分子量(MW)和溶解性存在差异一样,但二者在降低胆固醇效果上基本相同。

从燕麦或大麦中提取的天然的β-葡聚糖有个明显的特点就是分子量高(≥1 000 ku)。由于高分子量的β-葡聚糖应用于食品中会产生高黏性和不愉快的口感,需要生产低分子量的β-葡聚糖。通过大麦育种获得低黏度的β-葡聚糖不会影响其降低胆固醇的效果。另外,可通过添加外源性β-葡聚糖酶或激活内源性β-葡聚糖酶来制备低分子量的β-葡聚糖,该糖可能有更好的感官品质。但这些研究报道了酶水解高分子量β-葡聚糖会降低或丧失其降胆固醇的活性。然而,β-葡聚糖的分子量高低与降胆固醇效果间的关系难以确定,因为在这些研究中没有检测酶解前后β-葡聚糖分子量的高低。最近的研究结果认为中等分子量的β-葡聚糖具有与高分子量β-葡聚糖相似的降胆固醇效果。因此,当讨论食用高分子量的β-葡聚糖能降低血浆胆固醇浓度和食用低分子量的β-葡聚糖酶无此效果的时候,并不能确定来自谷物和酶降解的β-葡聚糖间分子量高低与降胆固醇效果是否有关。

本研究的目的是评价低分子量和高分子量的β-葡聚糖降低胆固醇的效果。高分子量的β-葡聚糖(MW = 1 000 ku)从大麦中提取,低分子量的β-葡聚糖(MW = 175 ku)是用从相同大麦中提取的β-葡聚糖经β-葡聚糖酶水解后获得。实验研究了低分子量和高分子量的β-葡聚糖对喂食高胆固醇饲料大鼠的脂蛋白胆固醇、血脂、粪便中中性固醇和大动脉中胆固醇沉积的影响。

第2课 Optimization of the jet steam instantizing process of commercial maltodextrins powders

1.原文

选自: Journal of Food Engineering, 2008,86(3): 444-452

作者: Takeiti C Y, Kieckbusch T G, Collares-Queiroz F P

Food powders are usually dissolved or dispersed in liquids before consumption but the moistening of fine particles is not straightforward due to the high interfacial tension, and therefore, such products are frequently instantized to enhance the wettability and the rate of dispersion/dissolution. Agglomeration is the most common technique used to produce instant food products. A proper enlargement of the particle should provide sufficient porosity in the agglomerates for rapid incorporation of liquid by capillarity (increase in wettability), an enhancement of the submergiblity of the powder inside the liquid followed by a fast dispersion of the powder and, if the substance is soluble in the liquid, the dissolution of the particles in it. The size of the agglomerates usually ranges between 0.2 and 2.0 mm and the product becomes more attractive in terms of color, particle shape and appearance than the original powder.

In addition to the reconstitution attributes the granules should present sufficient mechanical resistance to bear manipulation and transport. Unfortunately, it is difficult to improve all the characteristics of the powder simultaneously.[1] Enhancing the mechanical resistance of the agglomerates results in most cases, in the decline of the instant properties. The mechanism of the agglomerate formation depends on the feed formulation and conditions used. Turchiuli et al. (2005) reported[2] that agglomerates[3] obtained in a fluidized bed agglomeration process exhibit different shapes, structures and therefore[4] end-use properties depending on the type of binder used, its amounts and the way it is introduced in the equipment.

Despite being a research subject for years, there is no formal methodology for the development of an agglomeration process. Two approaches need to be considered: the prediction of the performance of the agglomeration process for new formulations and the adjustment of operation conditions in order to provide the desired improvement in the attributes of the product. Any desired and undesired agglomeration of amorphous substances is dependent on the mechanical properties of the entire particle or the particle surface. Changes in the mechanical properties of the material are linked to changes in moisture and temperature and can be predicted by applying the glass transition concept.

Maltodextrins have well defined physical properties, and, unlike natural starches, maltodextrins are soluble in water,[5] which popularized their use as additive in the food industry. They are applied, for example, as spray-drying aids for flavors and seasonings, carriers for synthetic sweeteners, flavors enhancers, fat replacers, and bulking agents.[6] Drying of sugar-rich fruits pulp into powder is difficult, mainly due to the presence of low molecular weight sugars, such as fructose, glucose, sucrose and acids present. Due to their low molecular weights, the molecular mobility of the materials is high when the temperature is just above the glass transition temperature (T_g). To overcome these problems, addition of dryings aids such as maltodextrin with different dextrose-equivalent (DE) are added to get nonstick and free flowing powder. In addition, maltodextrins were also used as food model since they present a wide distribution of molecular mass between oligosaccharides and polysaccharides. The objective of this work was to evaluate the process of steam jet agglomeration, by following a sequential strategy of experimental designs in order to, initially, identify the most important variables and, subsequently, optimize the conditions of the process which would result in a lower dissolution time.

2.词或词组

agglomeration *n.* 结块,凝聚,造粒

amorphous *adj.* 不定型的,无组织的

dextrose-equivalent(DE) 糖化度

fluidized bed 流化床

interfacial tension 界面张力

instantize *vt.* 把(食品、饮料)预先配制好,使调制快速方便化

maltodextrin *n.* 糊精

oligosaccharide *n.* 低聚糖,寡糖

polysaccharide *n.* 多糖

seasoning *n.* 调味品,调料

straightforward *adj.* 正直的,坦率的,简单的,易懂的,直截了当的; *adv.* 坦率地

submergibility *n*. 沉降性

glass transition temperature (T_g) 玻璃化转变温度

wettability *n*. 可湿性

3.主要语法现象

[1][6] 现在分词短语作主语

[2] 宾语从句

[3] 过去分词短语作主语

[4] 独立结构

[5] 定语从句

4.原文的参考译文

第2课 商品糊精粉的喷流预处理工艺的优化

食物粉末在使用前通常要溶解或分散,但由于细粒有很高的表面张力,不易直接湿润。因此,这种产品通常需经过增溶处理,以提高产品的可湿性和分散溶解速率。造粒是生产速溶食品最常用的技术,在造粒过程中,适当地增加粒度可赋予颗粒丰富的孔洞,利于通过毛细管迅速与水结合,加快粉末在水中沉降并快速分散,对于可溶性产品则可增加颗粒的溶解速度。粒度大小通常为0.2～2.0 mm,造粒产品比原粉末有更好的色泽、粒性和外观。

造粒的另一好处就是颗粒表现出良好的承受加工和运输的机械性能。遗憾的是要同时改善粉末的所有品质是困难的,增加颗粒的机械抵抗性会产生许多不利影响,如降低产品的速溶能力。颗粒形成时的机械性能取决于产品配方和加工条件。Turchiuli 等(2005)报道用流化床造粒制得的颗粒表现不同的外形结构,因此最终性能取决于所用输送带的类型、总量和入料方式。

尽管此课题研究了许多年,还没有形成正式的关于造粒工艺的方法论。需要考虑两种方法:对造粒过程产生的新结构的预测和为获得理想产品而对操作条件的调节。粉末状物质的造粒是否理想依赖于最终颗粒的机械特性和表面性能。材料的机械性能与水分和温度的改变相关,可用玻璃相变理论进行预测。

糊精被认为有良好的物理性能,不像淀粉,糊精可溶于水,因此广泛用作食品工业添加剂。例如,糊精可被用作香料和调味料的喷干助剂,用作合成甜味剂、风味强化剂、脂肪替代物和罐头食品添加剂。高糖果浆干燥成粉困难,主要是由于果糖、葡萄糖、蔗糖和酸等小分子物质的存在。因为分子量低,这些成分在温度高于玻璃化转变温度时分子活动性高。为了解决这些问题,可添加不同糖化度的糊精作为干燥助剂以获得非黏性的流动性好的粉末。另外,糊精用作食品原料是因为其有很宽的分子量范围,从寡糖到多糖。本研究的目的是评价气流造粒的生产工艺,遵循实验设计的顺序和策略,首先确定最重要的变量,然后优化过程条件,从而减少溶解时间。

第3课 Sterilization solutions for aseptic processing using a continuous flow microwave system

1.原文

选自：Journal of Food Engineering，2008，85(4)：528-536

作者：Coronel P，Simunovic J，Sandeep K P，Cartwright G D，Kumar P

Advanced and emerging thermal processing methods include pasteurization and sterilization using continuous flow microwave heating，in-package microwave processing，ohmic heating，radio frequency heating，infrared heating as well as other techniques and combinations of traditional and advanced processing methods of heating and cooling. Continuous flow microwave heating is one of these technologies in the food industry with a potential to replace the conventional heating process for viscous and pumpable food products. In contrast to conventional heating，heating by microwave provides volumetric heating of the entire food product. Continuous flow microwave heating is also associated with improved color，flavor，texture，and nutrient retention. Heating of food products[1] using continuous flow microwave systems has been reported by various researchers. Thus，aseptic processing using a continuous flow microwave system can combine the advantages of an aseptic process along with those of microwave heating.

Aseptic processing of foods requires[2] that both the processing and the packaging systems be free of microbiological contamination. A sterilization step，prior to processing，should be carried out to ensure that there is no contamination in the system. When conventional heat exchangers are used，it is a general practice in the industry to recirculate water[3] heated to 121 ℃ for a period of 30 min[4] after the farthest point of the system has reached that temperature.

However，the conventional method of sterilization by recirculating hot water cannot be applied to a continuous flow microwave heating system because the food product to be processed absorbs microwave energy in a very different manner as compared to water[5] depending on dielectric properties. Dielectric properties determine the extent of heating of a material when subjected to a microwave field. Therefore，knowledge of dielectric properties is important for the design of a continuous flow microwave heating system. Dielectric properties consist of dielectric constant (ε') and dielectric loss factor (ε''). Dielectric constant is a measure of the ability of a material to store electromagnetic energy whereas dielectric loss factor is a measure of the ability of a material to convert electromagnetic energy to heat. As a result，switch-over from water to the food product could result in under- or over-heating. Under-heating can lead to loss of sterility，rendering the process inadequate，while over-heating can potentially lead to runaway heating，flashing，or product deposition on the walls.

Therefore, a model solution with dielectric (dielectric constant, ε′ and dielectric loss factor, ε″) and flow properties similar to that of the food product is required as a sterilization solution. This solution will be heated[6] using microwave energy and recirculated through the processing system during sterilization. The matching of dielectric and flow properties is very important to avoid over- or under-heating in transition from sterilization solution to product.

This study deals with a method to prepare model solutions of liquid foods[7] that can be used as sterilization solutions for aseptic processing of foods using a continuous flow microwave heating system. The model solution will be prepared by matching the dielectric and flow properties of the solutions[8] made with simple solutes (table salt, sugar, and CMC) to those of the food product to be processed.

2. 词或词组

pasteurization *n.* 巴氏杀菌
sterilization *n.* 灭菌,杀菌,绝育
microwave *n.* 微波
ohmic *adj.* 欧姆的
infrared heating 远红外加热
pumpable *adj.* 可用泵抽吸(或抽送)的
retention *n.* 保持力
aseptic processing 无菌生产
recirculate *v.* 再流通,再循环
switch-over 转变
dielectric property 介电特性
dielectric constant (ε′) 介电常数
dielectric loss factor (ε″) 介电损失因子
electromagnetic energy 电磁能

3. 主要语法现象

[1] 现在分词作定语
[2] 宾语从句
[3][8] 过去分词作定语
[4] 状语从句
[5][6] 现在分词作状语
[7] 定语从句

4. 原文的参考译文

第 3 课 用于连续微波处理系统消毒的杀菌液

先进的和新型的热处理方法包括巴氏杀菌和连续微波加热、包装微波处理、电阻加热、射

频加热、红外加热以及其他方法的杀菌手段，还包括传统和先进相结合的加热和冷却方法。连续微波加热可用于食品工业中热处理高黏性和需泵输送的食品，可替代传统的热处理方式。相对于传统的热处理，微波加热可使整个产品同时受热升温，连续微波加热还可改善产品的色泽、风味、结构和营养保留。许多研究者报道了用连续微波设备加热处理食品的研究。因此，用连续微波系统灭菌处理可获得与那些微波加热同样的杀菌效果。

然而，采用循环热水的常规灭菌方法不能被运用于连续微波加热系统，因为按介电特性，被处理的食品吸收微波能的方式完全不同于水。介电特性决定了材料在微波范围中产热的程度。因此，介电特性的知识对设计连续微波加热系统是重要的。介电特性包括介电常数(ε′)和介电损失因子(ε″)。介电常数反映材料储存电磁能的能力，而介电损失因子反映材料将电磁能转化成热能的能力。结果显示，从水到食品的转换能导致过冷或过热。而过冷可能导致灭菌无效，生产不正常。而过热又会导致热量损失，闪光或产品沉积在壁上。

食品灭菌处理要求生产和包装线均不能受微生物污染。在生产前，应首先对生产线进行灭菌处理以确保整条生产线不受细菌污染。当使用常规的热交换机时，工厂通常的操作是用加热到 121 ℃的循环水使整个系统达到杀菌温度后保持 30 min。

因此，需要一种介电常数(介电常数 ε′和介电损耗因子 ε″)和流动特性类似于食品的模型溶液作为灭菌溶液。该溶液在灭菌过程中用微波能加热，并在生产系统中循环。相称的介电特性和流动性对避免从杀菌溶液到产品的交换过程中出现过热或过冷现象是非常重要的。

本研究提供了一种可用于进行食品无菌生产的连续微波加热系统消毒的，类似液态食品的杀菌溶液的制备方法。这种由简单溶质(食盐、糖以及羧甲基纤维素)配成的杀菌溶液的介电特性和流动特性与生产的食品相匹配。

第4课 Effect of fermentation temperature and culture media on the yeast lipid composition and wine volatile compounds

1. 原文

选自： International Journal of Food Microbiology，2008，121(2)：169-177

作者： Beltran G，Novo M，Guillamón J M，Mas A，Rozès N

Low temperature alcoholic fermentations are becoming more frequent due to the winemaker's tendency to produce wines with more pronounced aromatic profile. Wines produced at low temperatures (10-15 ℃) develop improved characteristics of taste and aroma. The temperature of fermentation affects both the retention of some varietal compounds and the production of fermentative metabolites. The improved quality of wines[1] produced at low temperatures can be attributed to a greater retention of terpenes, a reduction in higher alcohols and an increase in the proportion of ethyl and acetate esters in the total volatile compounds. The complexity of the fermentative medium also influences the yeast metabolism and thus, the fermentation performance and the final quality of the wine. Synthetic media[2] simulating grape must are widely used in laboratory fermentations to keep a constant composition for the study of wine fermentation and wine yeast metabolism. However, wines produced from natural grape must have some varietal compounds and precursors,[3] which could be incorporated to yeasts and modify some aspects of its metabolism and physiology, and consequently, the final wine composition. These changes can be more pronounced in specific conditions, such as fermentations performed at lower temperatures.

Although low temperature fermentation has interesting applications in the enological industry, this practice also has some disadvantages. The optimal growth temperature for *Saccharomyces cerevisiae* is 25 ℃, whereas 13 ℃ is a restrictive temperature that increases the risk of stuck or sluggish fermentations as reviewed by Bisson (1999). In these circumstances, the survival of cells can depend on their ability to adapt quickly to the changing environment. Changes in plasma membrane composition may be an adaptative response by the yeast since it is highly variable and clearly influenced by environmental factors such as temperature, oxygen, nutrient limitation and growth rate. These changes in plasma membrane are mainly accounted for by modifications in the lipid composition.

The main lipid components of eukaryotic membranes are phospholipids, sterols, sphingolipids and glycerolipids. Alterations in fatty acid, phospholipid and sterol levels are needed to maintain ethanol tolerance. The membrane fatty-acyl composition of yeast, like that of many other microorganisms, also changes with temperature: the lower the temperature, the more unsaturated the membrane fatty-acyl composition. However, the fatty acid composition of a cell can also be influenced by the environment's lipid composition since it can include fatty acids from the medium in its own phospholipids.

In grapes, unsaturated fatty acids (UFA) represent the main lipid component. The most abundant of these is linoleic acid, followed by oleic, linolenic and palmitoleic acids. Of the saturated fatty acids (SFA), palmitic acid is the most abundant. However, the initial fatty acid content of the must depends on which technological procedures, such as pressing, maceration or clarification, are applied.

[4] Regarding sterol content of natural musts, the main phytosterol in these musts was β-sitosterol. In white wine production, the absence of oxygen suppresses fatty acid desaturation and the sterol biosynthesis by yeast,[5] which reduces the capacity to synthesize unsaturated fatty acids and ergosterol, both essential for protecting the yeast against ethanol stress. In these circumstances, yeasts can incorporate exogenous sterols and unsaturated fatty acids, as it might be the case in industrial fermentations. On the other hand the reduced oxygen presence during the fermentation induces the synthesis of medium-chain fatty acid (C6 to C14) and their corresponding ethyl esters, which are toxic to the cells.

The fermentation temperature and the culture media affect therefore the lipid metabolism,[6] which is related to cell development, membrane integrity and the production of several by-products, especially those directly related to wine aroma. The aim of this study is to analyse the lipid metabolism response of *S. cerevisiae* fermenting in different media: one without lipids such as the synthetic must and another with the presence of lipids (grape must),[7] that can be incorporated into the yeast membrane. During fermentation lipid metabolism is impaired and, thus, the response in the synthetic media will reflect the yeast metabolic response by itself. Furthermore, these differences between the two media are analysed at lower temperature (13 ℃),[8] where the response in lipid metabolism is highly needed for the cells to adapt their membrane fluidity. Also, the effects of both conditions (media and temperature) upon wine volatile compounds closely related to lipid metabolism and that affect strongly wine quality are analysed.

2. 词或词组

aromatic *adj*. 芳香的
aroma *n*. 芳香，香气，香味
terpene *n*. 萜烯,萜(烃)
ethyl ester 乙酯
metabolism *n*. 新陈代谢
media *n*. 培养基,媒体,媒介
grape must 葡萄果汁
varietal *adj*. 品种的
enological *adj*. 葡萄酒酿造的
sluggish *adj*. 行动迟缓的
sphingolipid *n*. 磷脂
glycerolipid *n*. 甘油脂质
maceration *n*. 浸泡,泡软
clarification *n*. 澄清,净化
linoleic acid *n*. 亚油酸
oleic acid *n*. 油酸
linolenic acid *n*. 亚麻酸
palmitoleic acid *n*. 棕榈油酸
must *n*. 待发酵的葡萄汁;*aux*. 应该,应当,必须,一定
phytosterol *n*. 植物甾醇类
β-sitosterol *n*. 谷甾醇
ergosterol *n*. 麦角固醇

3. 主要语法现象

[1] 过去分词短语作定语
[2] 现在分词作定语
[3][5][6][7][8] 非限制性定语从句
[4] 独立结构

4.原文的参考译文

第4课 发酵温度和培养基对酵母脂肪组成和葡萄酒香气组成的影响

低温酒精发酵变得更普遍,因为酿酒商倾向于生产更具有显著芳香特征的葡萄酒。低温(10～15℃)条件下生产的葡萄酒,其口味和香气均得以改善。发酵温度对一些具有原料特征的化合物的保留和发酵代谢产物的产生均有影响。低温生产的葡萄酒的品质提高表现在极大地保留了萜烯类物质、减少了高级醇和增加乙酯和乙酸酯在总挥发物中的比例。发酵培养基的复杂性也会影响酵母的代谢,进而影响发酵的完成和葡萄酒成品的品质。在研究葡萄酒发酵和葡萄酒酵母代谢的实验室发酵中,模拟葡萄汁的合成培养基被广泛地应用,以保持成分的稳定。但天然葡萄汁酿造葡萄酒含有一些特有化学成分和前体,它们会同酵母结合,改变酵母的代谢和生理特性,最终改变产品组成。在特定条件下,这种改变会更加明显,如在低温发酵时。

虽然,低温发酵已在葡萄酒工业中应用,但在实际应用中还存在一些问题。酿酒酵母(*Saccharomyces cerevisiae*)的最佳生长温度是25℃,但13℃是其限制温度,在此温度下会增加发酵缓慢或停滞的风险性。低温环境下,细胞的生存取决于其对迅速变化的环境的适应能力。细胞膜组成的改变是酵母适应性的反应,因为它有很高的可变性,且明显受温度、氧气、营养局限性和生长速率等环境因素的影响。细胞膜的改变主要是脂质组成的变化引起的。

真核细胞的细胞膜主要脂肪成分是磷脂、甾醇、(神经)鞘脂类和甘油脂质。脂肪酸、磷脂和甾醇水平的改变会影响其对乙醇的耐受力。像许多其他微生物一样,酵母的膜脂肪酰基成分也常随温度变化:温度越低,膜脂肪酰基成分的不饱和度越高。然而,细胞的脂肪酸组成也受环境中脂肪组成的影响,因其磷脂中含有来自培养基的脂肪酸。

葡萄中不饱和脂肪酸是其脂肪的主体,其中最丰富的是亚油酸,其次是油酸、亚麻酸和棕榈油酸。饱和脂肪酸棕榈酸的含量最丰富。然而,葡萄汁中初始的脂肪酸含量取决于使用的生产工艺,如压榨、浸泡或净化等。

天然葡萄汁的甾醇含量,其中主要的植物甾醇类是β-谷甾醇。在白葡萄酒生产中,缺乏氧气会抑制酵母对脂肪酸的不饱和化和甾醇生物合成,从而降低不饱和脂肪酸和麦角甾醇的合成能力,而二者是保护酵母抵抗乙醇胁迫所必需的。在此环境中,酵母可能与外源性甾醇和不饱和脂肪酸结合,这可能在工业发酵中出现。在另一方面,在发酵中降低氧压会诱导中链脂肪酸(C6～C14)及其对应的乙酯的合成,乙酯对细胞有毒性。

因此,发酵温度和培养基会影响类脂代谢,脂代谢与细胞生长、膜完整性和一些直接影响葡萄酒风味的副产物产生相关。本研究的目标是分析在不同培养基下,与酿酒酵母(*S. cerevisiae*)发酵相关的脂肪代谢:一种培养基是不含脂质的合成果汁,另一种培养基是含有可以被结合到酵母膜中的脂质(葡萄汁)。在发酵时脂质代谢被削弱,因而在合成的培养基中的反应是酵母自身新陈代谢的反应。此外,分析了两种媒介在低温(13℃)下的差异性,脂质代谢反应是反映细胞对膜流动性的适应能力所需要的。同时还分析了两个条件(培养基和温度)对与脂质代谢紧密相关的和直接关系葡萄酒品质的挥发性成分的影响。

第5课 Development and assessment of pilot food safety educational materials and training strategies for Hispanic workers in the mushroom industry using the Health Action Model

1.原文

选自：Food Control，2008，19(6)：616-633

作者：Nieto-Montenegro S，Brown J L，LaBorde L F

Worker[1] mishandling of food is one of the major causes of foodborne disease outbreaks. Because outbreaks often lead to severe economic losses and even bankruptcy, food handler training is an important business strategy for managing food safety risks. However, there is no evidence[2] that worker practices improve when training programs provide only information. Several studies have demonstrated that[3] increasing knowledge does not necessarily lead to changes in behaviors. To be effective, training programs should be based on appropriate adult education theory. They should incorporate activities[4] that support skills development relevant to real life situations[5] in which the workers can put information into practice. For example, a training session[6] that raises awareness of the possibility[7] that *E. coli* bacteria may accumulate under fingernails should also demonstrate the correct handwashing procedure and require the learner to practice until he or she can successfully demonstrate that procedure. Such training sessions then become active learning sessions. Adequate resources and a receptive management culture also are fundamental to applying good food safety practices. These authors believe that an effective food safety program should be relevant to the situation, teach skills, promote active learning, increase risk perception and consider the work environment. Program evaluation is also a critical part of a food safety-training program since it allows the implementers to assess its effectiveness.

There are few studies of how training programs impact food safety behaviors within food production and processing settings. Most of these studies describe the program implementation or training strategies utilized, but do not focus on program evaluation. In contrast, several studies[8] conducted within the foodservice and hospitality industries have evaluated pre- and immediate post-intervention knowledge and self-reported behaviors among workers[9] who have attended food safety workshops.

For effective food safety programs, evaluation must be based on careful planning. Several studies have concluded that[10] observing workers during day-to-day operations is important for planning and evaluating training programs. Studies to assess consumer food safety behaviors have similarly used direct observations of food handlers. These studies often find that self-reported behaviors obtained through surveys do not always match behaviors obtained through direct observations. Direct observations should provide more accurate and reliable assessment of actual food safety practices in a food industry setting. In addition, good programs need a sound theoretical framework.

2. 词或词组

foodborne　*n.*　食物，养料

bankruptcy　*n.*　破产

strategy　*n.*　策略

fundamental　*adj.*　基础的，基本；　*n.*　基本原则，基本原理

awareness　*n.*　知道，晓得

perception　*n.*　理解，感知

3. 主要语法现象

[1][3] 现在分词作定语

[2][4][6][7][9] 定语从句

[5] 状语从句

[8] 过去分词作定语

[10] 现在分词作主语

4. 原文的参考译文

第5课　开发与评估食品安全教育的材料，以及对蘑菇厂的西班牙工人进行健康行为规范的培训策略

工人在食品生产中不规范操作是食品传染病暴发的主要原因。由于暴发经常导致严重的经济损失甚至破产，所以对食品管理者的培训是防范食品安全性风险的一项重要经营战略。然而，若培训计划仅停留在书面，工人的实际操作则不会有所改善。几项研究表明，增长的知识不一定能获得行为上的改变。为了使培训有效，培训方案应该以适当的成人教育理论为基础。他们应该与实践活动及技能培训相结合，有利于工人将理论变为实践。例如，在提高对大肠杆菌（*E. coli*）在指甲盖聚集可能性的认识培训会上，也应该示范正确洗手的过程，并要求每位参训人员当场练习，直到能正确按示范操作。这种培训就变为一种积极主动的培训。充足的经费和人性化的管理文化也是实施良好食品安全操作的基础。作者相信有效的食品安全计划应结合实际，教授技能，促进主动学习，增加风险意识，分析工作环境。计划评估能让执行者评价计划的有效性，因此计划评估也是食物安全训练计划的一个重要部分。

很少有研究培训项目如何影响食品生产和加工环境中的食品安全行为。这些研究大多描述了所使用的项目实施或培训策略，但不侧重于项目评估。相比之下，在食品服务和酒店业内进行的几项研究评估了参加过食品安全培训班的工人在干预前期、中期和后期的认知和自我监督行为。

有效的食品安全计划，评估必须根据周密的计划进行。几项研究认为，在日常操作中观

察工人对于规划和评估培训计划很重要。评估消费者食品安全行为的研究也同样使用了对食品处理人员的直接观察。这些研究经常发现,通过调查获得的自我报告的行为并不总是与通过直接观察获得的行为相符。在食品加工的环境中,直接观察能对实际食品安全操作提供更准确和可靠的评估。此外,好的计划需要一个健全的理论框架。

第6课 Determination of O_2 and CO_2 transmission rates through microperforated films for modified atmosphere packaging of fresh fruits and vegetables

1.原文

选自:Journal of Food Engineering, 2008, 86(2): 194-201

作者:González J, Ferrer A, Oria R, Salvador M L

Microperforated films are commonly used in the modified atmosphere packaging (MAP) of high respiration fresh food products; for example minimally processed fruits and vegetables. The perforations of a plastic container act like a polymeric film in the regulation of gas exchange, although there are marked differences compared with conventional films. These perforations allow a much higher exchange of gases than the said conventional films. The diffusion of O_2 and CO_2 through air is 8.5 and 1.5 million times greater, respectively, than through low density polyethylene films. This difference in gas diffusion means[1] that the gas exchange of a container occurs almost entirely through the microperforations in relatively impermeable films. The other gas exchange differences lie between the ratio of the permeability for CO_2 and O_2; close to a ratio of 1 in microperforated films. The ratio,[2] which is commonly denoted as β, is very different for polymeric films, being usually between 3 and 6. A relatively high concentration of CO_2 is not reached in containers with non-perforated films. This is taking into account the respiration coefficient of the packaged product[3] which can fluctuate, if there is no temperature abuse, between 0.7 and 1.3.

One of the first steps in the design of a MAP container is to predict the permeability necessary for both CO_2 and O_2. The permeability of conventional films can be measured with a diffusion cell,[4] following one of the standard dynamic analysis methods (for example ASTMD 3985 for oxygen and ASTMD 1343 for carbon dioxide). However, most experimental systems for measuring the permeability of perforated or microperforated plastics are static describe a flow-through method for measuring permeability, with adequate control of the pressure at both sides of the microperforation. Nevertheless, dynamic methods are of doubtful application due to the gas convection that takes place when the pressures between the two cells are even just slightly unbalanced.

Various models have been proposed[5] which attempt to describe the exchange of gases through perforation. Some authors have used the flow equation according to the Stephan-Maxwell law. Chung et al. (2003) consider the application of the Stephan-Maxwell law to be more accurate but less versatile because of its complexity, and therefore prefer to apply Fick's law to estimate the flow of small leaks.

The application of Fick's law is much more widespread since it enables one to work with simple and reasonably representative expressions. Thus, Emond et al. (1991) and Fonseca et al. (1996) used it to predict the gas exchange rate through perforations. They established by means of empirical equations the gas exchange relationship with the diameter of the perforation, thickness of the film and even the temperature. Fishman et al. (1996) regarded gas transport through perforations as a problem of macroscopic diffusion in a cylindrical pathway filled with air; exclusively obeying Fick's law. They considered the total diffusive pass length of a perforation as the sum of the perforation length and end correction. This correction term was proposed by Heiss (1954) and more recently by Lange et al. (2000). Its value ranges between 5/12 and 10/12 times the perforation diameter,[6] depending on the air speed outside the leak. The need for an end correction term arises from the fact[7] that the gas around the leak ends may or may not be in equilibrium with the surrounding bulk gas.

A substantial proportion of publications in the field refer to quite large perforations in their work (from up to 11 mm in diameter). However, the microperforations are small holes in film from 50 to 200 μm in diameter. The extrapolation results[8] obtained with relatively large perforations to small perforations can give rise to errors; and it should be remembered[9] that the design of these containers requires a high degree of accuracy. Inadequate design can be ineffective or even harmful to maintaining the quality of the product.

The few existing studies[10] dealing with microperforated films relate permeability to the density of the perforations in a container. Other studies relate the orientation of the container and the position of the perforation or perforations in a container with the gas composition inside. The influence of the dimensions of the microperforation on the transmission rate is dealt with in very few works, such as Ghosh and Anantheswaran (2001). This situation exists despite the fact[11] that this type of perforation, with a diameter of less than 200 μm, is extensively used in MAP.

This current work uses a static method to obtain experimentally the evolution over time of the O_2 and CO_2 concentration in the interior in an impermeable microperforated container. The oxygen and carbon dioxide transmission rates are determined from the experimental data by applying Fick's law. Several plastics have been used, microperforated by laser,[12] whose characteristics in terms of thickness and perforation size are very common in fresh-cut vegetable and fruit packaging. The O_2 and CO_2 transmission rates obtained for the 29 perforations studied are related with various parameters: perforation size, total exchange area, equivalent radius and film thickness.

2.词或词组

microperforate *v*. 微穿孔
the modified atmosphere packaging (MAP) 气调包装
respiration *n*. 呼吸,呼吸作用
polymeric films 聚合膜
permeability *n*. 渗透性
dynamic *n*. 动力学的,动态的
transmission *n*. 播送,发射,传动,传送,传输,转播
equivalent radius 相同半径
impermeable *adj*. 不能渗透的
interior *adj*. 内部的,内的; *n*. 内部

3.主要语法现象

[1] 宾语从句
[2] 补语从句
[3][5][7] 定语从句
[4][6] 现在分词作状语
[8] 过去分词作定语
[9] 主语从句
[10] 现在分词作定语
[11] 定语从句
[12] 非限制性定语从句

4.原文的参考译文

第6课 用于新鲜蔬菜和水果气调包装的微孔膜的 O_2 和 CO_2 透过率的测定

微孔膜常用于高呼吸作用的新鲜食品产品的气调包装(储藏),例如用于低加工度的水果和蔬菜。塑料容器的穿孔在调节气体交换方面的作用类似于聚合物薄膜,尽管与常规薄膜相比有显著差异。这些穿孔比上述传统的膜有更多的气体交换。O_2 和 CO_2 在空气中的扩散分别是低密度聚乙烯薄膜的 8.5×10^6 倍和 1.5×10^6 倍。在气体扩散上,这一差异意味着容器的气体交换几乎完全地通过相对没有渗透性的膜上的微孔。气体交换的其他差异在于 CO_2 和 O_2 的渗透率,微孔膜的比率接近1。这一比率通常用 β 表示,对于聚合膜来说是非常不同的,通常为3～6。在非穿孔性膜容器中 CO_2 很难达到较高的浓度。也应考虑包装的鲜活产品的呼吸作用系数,如果没有不适的温度,通常为0.7～1.3。

MAP容器设计首先应预测 CO_2 和 O_2 的渗透性。常规膜的渗透性可以根据一个标准动态分析方法(如氧气按ASTMD 3985,二氧化碳按ASTMD 1343),用扩散室法测量。然而,多数测量穿孔或微孔塑料渗透性的实验系统是静态的描述用于测量的渗透性的流通方法,对微孔

两边压力适当控制。不过，动态方法的应用是不可靠的，因为在两室间细小的压差均会引起气体对流。

多种不同的描述气体交换的模型被提出。有些作者根据 Stephan-Maxwell 定律建立了流体运动方程。Chung 等(2003)认为 Stephan-Maxwell 法则的应用更加准确，但是由于太复杂，不能通用，因此宁愿用 Fick's 法则评估微细的泄漏流量。

Fick's 法则的应用更加普遍，因为它能用简单、相当典型的公式表示。因此，Emond 等(1991)和 Fonseca 等(1996)用它预测通过孔洞的气体交换率。他们通过经验方程式建立了气体交换与孔径、膜厚，甚至温度间的关系。Fishman (1996)把通过孔的气体流动看成在充满空气的圆柱体中大规模扩散的问题，专一性地遵从 Fick's 法则。他们将孔的总扩散路径考虑为孔长和最终修正值的总和。这种修正值由 Heiss (1954)提出，而最近 Lange 等 (2000)提得最多。这种修正值是孔径的 5/12～10/12，大小取决于由缝隙向外泄漏的气流速度。最终修正值会随缝隙处气体与周围容器中游离气体平衡与否而变化。

在此领域发表的大部分研究中都采用的是大孔 (最大直径达 11 mm)。然而，微孔是膜上孔径为 50～200 μm 的小孔。将从较大的孔获得的结果推测到小孔会产生错误。还必须记住对这些容器的设计要求高度准确。不恰当的设计可能是无效的，甚至对产品品质的保持造成危害。

现有的少数涉及微孔膜的研究将渗透率同容器中孔密度联系起来。其他研究将容器方位和容器上孔的位置与内部气体成分联系起来。极少人研究微孔大小对气体传输率的影响，如 Ghosh 和 Anantheswaran (2001)。尽管这种孔径小于 200 μm 的孔被广泛地用于 MAP，但这种影响气体传输率的情况仍存在。

本研究用静态法获得在无渗透性的经微孔处理的容器内部 O_2 和 CO_2 浓度随时间的变化。运用 Fick's 法则从实验性数据中确定 O_2 和 CO_2 传输率。实验使用经激光微孔处理的几种塑料，塑料的厚度和穿孔大小在新鲜蔬菜和水果的包装中非常普遍。研究了制备的 29 种孔的 O_2 和 CO_2 通透率与各种参数的相关性，包括：孔大小、总交换区域、等效半径和膜厚度。

作　业　汉译英

1. 谷物中天然的混合的结合 β-glucan 被划分为可溶性膳食纤维，与瓜尔胶和其他任意卷曲的多糖有相似生理学效果。燕麦和大麦产品能削弱餐后血糖和胰岛素反应的能力，这与其所含的(1，3)(1，4)-*β*-*D*-葡聚糖 (*β*-葡聚糖)和黏度相关。*β*-葡聚糖的黏度对降低血清胆固醇水平的作用没有被直接证明，且并不是所有研究报道都有统计学上的显著降低。

2. 糊精，一种部分水解淀粉，由支链和直链分子组成的混合物，二者对吸湿性和成片特性(如张力强度、易碎性和崩解时间)都有影响。因此，本报道描述了糊精等级(不同支/直链比率)对粉末吸湿性、流动性、密度和压实度的影响。

3. 连续微波巴氏杀菌系统给饮料工业带来许多好处，已在多种饮料中试验，如苹果汁、牛奶和橙汁。然而，许多对微波巴氏杀菌的研究只注重微生物和酶的失活，很少评价连续微波巴氏杀菌系统的过程和生产参数。因此，本课题的目的就是设计用于苹果酒的实验室型连续微波巴氏杀菌系统以及确定生产参数。

4.在美国,食源性疾病是绝对的公众健康问题,对食品生产者的教育和培训非常关键,因为,工人的错误操作会导致食源性疾病的暴发。现在,虽有许多食品安全教育资料可用,不幸的是,工人的文化、经济和社会背景的差异,使得千篇一律的食品安全教育计划难以实施。另外,如果教育材料设计不顾目标听众周围场地的社会、自然和环境因素,这些教育材料可能是无效的。

第6章

论文的材料与方法

Materials and Methods of Article

第1课 Hypobaric storage removes scald-related volatiles during the low temperature induction of superficial scald of apples

1.原文

引自： Postharvest Biology and Technology，2000，18:191-199

作者： Zhenyong Wang，David R. Dilley

Materials and methods

Plant materials

'Law Rome'，'Granny Smith'，'Red Delicious'，'Mutsu' and 'Golden Delicious' apples were harvested at the preclimacteric stage of maturity from the Michigan State University Clarkesville Experiment Station (CHES).

***Hypobaric studies* 1997/1998**

'Law Rome' and 'Granny Smith' fruit were stored hypobarically or in air at 1 ℃. Fruit were placed under hypobaric storage immediately after harvest or after 0.5，1，2，3，4，5 or 6 months storage in air at 1 ℃ to determine the effects of delaying imposition of hypobaric storage on ripening and scald development，and production of α-farnesene and MHO. After total 6 months of storage，fruit were removed from hypobaric or air storage and transferred to 20 ℃ in air in 4-l glass jars. α-Farnesene and MHO production rates at daily or bi-daily intervals were measured as described by Wang and Dilley (2000). After a 90-min enclosure，the headspace of a glass jar with about 1.2 kg of fruit was sampled by solid phase microextraction (SPME)[1] equilibrating the fibre for 4 min to absorb volatiles. The volatiles were measured by gas chromatography/mass spectrometry (GC/MS) as described by Song et al. (1997) to determine volatile production/evolution. At the same time intervals，14 mg of epicuticular wax sample was removed from five fruit in each treatment and placed in a 2-mL glass vial for 3 h at 20 ℃ to determine the amount of α-farnesene and MHO partitioned in the epicuticular wax by SPME/GC/MS as described above. Fruit in parallel samples were used for fruit firmness and ripening changes immediately upon removal from storage and again after 7 days at 20 ℃.

2.词或词组

climacteric stage　呼吸跃变期

epicuticula　*n*.　上表皮

α-farnesene　*n*.　α-金合欢烯

gas chromatography/mass spectrometry (GC/MS)　气相色谱/质谱

hypobaric storage　减压贮藏

MHO(6-methyl-5-hepten-2-one)　6-甲基-5-庚烯-2-酮，甲基庚烯酮

solid phase microextraction (SPME)　固相微萃取

volatile　*adj*.　可挥发的，易挥发的；　*n*.　挥发物

3.主要语法现象

[1] 现在分词短语作定语

4.原文的参考译文

第 1 课　减压贮藏可以去除低温导致苹果表皮烫伤过程中烫伤相关的挥发物

材料与方法

植物材料

"Law Rome""Granny Smith""Red Delicious""Mutsu""Golden Delicious" 5 种苹果在呼吸跃变前期采收于密歇根州立大学 Clarkesville 试验基地(CHES)。

减压贮藏试验(1997—1998 年度)

分别将"Law Rome"和"Granny Smith"减压贮藏或在常压 1 ℃条件下贮藏。果实分别在不同的条件下贮藏:采收后立刻置于减压条件下贮藏;采收后分别在 1 ℃条件下贮藏 0.5 个月、1 个月、2 个月、3 个月、4 个月、5 个月及 6 个月后再减压贮藏，由此来确定延迟减压贮藏对果实后熟、褐变发展及 α-金合欢烯和甲基庚烯酮生成的影响。在总计 6 个月的贮藏后，将果实从减压或 1 ℃空气中移到玻璃罐中，每个样品装 4 瓶，每瓶 1.2 kg，放置在 20 ℃条件下。每隔一天或两天采用 Wang 和 Dilley(2000)报道的方法分别测定 α-金合欢烯和甲基庚烯酮的产生速率。封闭 90 min 后，用固相微萃取法(SPME)取瓶顶隙的气体为检测样品，平衡 4 min 充分吸收挥发物。用气质联用(GC/MS)方法(Song 等，1997)测定挥发物的产生和变化。同时每个样品取 5 个果实，从果皮上取 14 mg 上表皮蜡质放于 2 mL 玻璃瓶中，在 20 ℃下放置 3 h，后用上述的(SPME/ GC/MS)方法测定表皮蜡质中释放的 α-金合欢烯和甲基庚烯酮量。其他的平行样用于测定贮藏后果实硬度和后熟变化，并在 20 ℃下放置 7 d 后再测一次。

第 2 课　Determination of phenolic compounds and their antioxidant activity in fruits and cereals

1.原文

选自：Talanta，2007，71：1741-1751

作者：Stratil P，Klejdus B，Kubáň V

Materials and methods

Instruments

A spectrophotometer HELIOS β controlled with a VISION 32 Software (Spectronic

Unicam, Cambridge, UK); an ultrasonic bath HD 2070 (model M8 72, Bandelion Sonoplus, Germany), a high-speed Grindomix mill (Retsch, Germany) and a CHRIST ALPHA 1-2 B lyophiliser (Braun Biotech International, Germany) were used for sample preparation.

Chemicals

Caffeic, ferulic and ascorbic acids (purity = 99.0% each), 2,2-diphenyl-1-picrylhydrazyl radical (DPPH≈90.0%) and 2,2′- azinobis(3-ethylbenzothiazolin-6-sulfonate) diammonium salts were purchased from Sigma-Aldrich Chem. Comp. (USA); gallic acid monohydrate (= 98.0%), Trolox (6-hydroxy-2,5,7,8-tetramethylchromane-2-carboxylic acid, a hydrophilic derivative of tocopherol, purum, =99%, for HPLC), Folin-Ciocalteu reagent (FC reagent) and 2,4,6-tris(2-pyridyl)-s-triazine (TPTZ, puriss, =99.0%) were obtained from Fluka Chemie (Buchs, Switzerland). Methanol and acetonitrile of gradient grade were purchased from Merck (Darmstadt, Germany). Other chemicals of p.a. purity were from Pliva-Lachema (Brno, Czech Republic). All reagents and standard solutions were prepared using Milli Q deionised water (Millipore, Bedford, USA).

Sample preparation

Samples (17 kinds of fruits and 6 kinds of grains mostly commonly consumed in Czech Republic) were purchased from a local market (Delvita stores). Less usual or seasonal fruits (sea-buckthorn and black rowanberry, apricots, peaches, plum, greengage and red current) were gained from own sources and kept frozen at − 20 ℃ until freeze-drying. Edible parts of fruits (20-50 g) were cut into small pieces, lyophilised and dry matter was determined gravimetrically (± 1 mg). Lyophilisates were homogenized in a laboratory ultra-mixer and the powder was stored in plastic bottles under nitrogen at − 20 ℃ until analysed. Cereals and products of them, such as flowers, were used in natural state without further dehydration. Vinson et al. procedure for extraction of free and total (conjugated) phenolic compounds was modified. Details of this procedure were published in a previous paper.

Determination of phenolic compounds in extracts

Performance of Folin-Ciocalteau method according Singleton et al. and Vinson et al. was described in details in our previous paper. PBM procedure was performed according to Waterman and Mole and AAPM according Schoonen and Sales. All three methods were modified to a reaction volume of 1 mL. Reactions and incubations were carried out in plastic test tubes and spectrophotometric measurements were made in cuvettes with 1 cm light pathway.

Determination of ascorbic acid

High-performance liquid chromatographic method with mass spectrometric detection (HPLC/MS) was selected for assessment of ascorbic acid concentration in extracts. Standard solution of ascorbic acid in aqueous methanol (1:1, *V/V*) was prepared. Ascorbic acid was determined using HP 1100 liquid chromatograph equipped with HP MSD 1100 (Hewlett-Packard), a Waters dC 18 chromatographic column (4.6 mm×20 mm, 3 μm) at the flow rate 1.5 mL/min and the column thermostat set at 20 ℃. Mobile phase consisted of A: 0.01% trifluoroacetic acid, B: acetonitrile (ACN); gradient: t = 0.0 min: 1% B; t = 0.2 min: 1% B; t = 3.0 min: 30% B; t = 3.5 min: 50% B; t = 5.0 min: 1% B. Specific MS detection: SIM mode (ESI negative), m/z 175.0; gain 1.0; fragmentor 70; drying gas 8.01/min, nebulizer pressure 60 psi; temperature of drying gas flow 300 ℃; capillary voltage 3 000 V.

2. 词或词组

acetonitrile　*n.*　乙腈

2,2′- azinobis(3-ethylbenzothiazolin-6-sulfonate) diammonium salts　2,2-吖嗪双(3-乙基-苯并噻唑啉-6-磺酸)二铵盐

black rowanberry　黑果花楸

chromatographic column　色谱柱

cuvette　*n.*　比色皿,小玻璃管

ferulic acid　阿魏酸

greengage　*n.*　青梅

gallic acid monohydrate　没食子酸水化物

homogenize　*v.*　均质

lyophilize　*v.*　冷冻干燥

nebulizer pressure　喷雾压力

red current　红醋栗

sea-buckthorn　*n.*　沙棘

spectrophotometer　*n.*　分光光度计

tocopherol　*n.*　生育酚

trifluoroacetic acid　三氟乙酸

3. 原文的参考译文

第 2 课　水果和谷物中酚类物质的测定及其抗氧化活性研究

材料和方法

设备

HELIOS β 分光光度计,装有 VISION 32 软件(Spectronic Unicam,剑桥,英国);HD2070 超声波水浴锅(M872 型,Bandelion Sonoplus, 德国)、高速粉碎碾磨机(Retsch,德国)和 CHRIST ALPHA1-2B 冷冻干燥机(Braun Biotech International,德国)用于样品前处理。

化学试剂

咖啡酸、阿魏酸、抗坏血酸(纯度均为 99.0%),2,2-二苯基-1-吡啶并肼基(DPPH≈90%),2-2-吖嗪双(3-乙基-苯并噻唑啉-6-磺酸)二铵盐均购于美国 Sigma-Aldrich 公学公司。一水没

食子酸(=98.0%);Trolox(6-hydroxy-2,5,7,8-tetramethylchromane-2-carboxylic acid,一种亲水的生育酚衍生物,色谱级,纯度为99%),Folin Ciocalteu 试剂(FC 试剂),三吡啶基三嗪(TPTZ,特纯级,纯度为99%)由 FLUKA 化学公司提供(Buchs,瑞士)。各梯度乙醇及乙腈购于 Merk 公司(Darmstadt,德国)。其他各种化合物均为分析纯,购于 Pliva-Lachema 公司(Brno,捷克共和国)。所有试剂和标准溶液都用 Milli Q 去离子水(Millipore, Bedford,美国)配制。

样品处理

17种水果和6种谷物大多为捷克共和国日常消费品种(表1)购于本地市场(Delvita 市场)。不常见或者季节性水果(沙棘、黑果花楸、杏、桃、李、青梅及红醋栗)由自己提供并置于氮气中维持在-20℃以供分析时使用。取果实可食部分20~50 g 切成小片,冻干并测干重(±1 mg)。用超混机将冻干物粉碎成粉状,然后转到塑料瓶中放置在-20℃条件下待测。谷物及其制品比如面粉不需要进一步干燥直接用于实验。游离酚及总酚的萃取参考 Vinson 等的方法并作适当改进,具体方法参见文献。

萃取物中酚类物质的测定

Folin-Ciocalteau 方法参照 Singletion 等和 Vinson 等的研究结果,并在作者以前的文章中有详细描述。PB 方法按照 Waterman 和 Mole 所报道的步骤,AAPM 参照 Schoonen 和 Sales 的描述。以上三种方法均将反应体积改为14 mL。反应和保温都在塑料试管中进行,用1 cm 比色皿进行比色。

抗坏血酸的测定

采用高效液相色谱-质谱法(HPLC/MS)测定萃取物中抗坏血酸的浓度。标准抗坏血酸溶液为1∶1(体积分数)的酒精溶液。抗坏血酸的测定用 HP1100/MSD 液质联用仪测定,色谱柱为 Waters 的 dC 18 柱(4.6 mm×20 mm, 3 μm),流速1.5 mL/min,柱温箱温度20℃。流动相 A:0.01%三氟乙酸,B:乙腈(ACN);梯度:t =0.0 min: 1% B; t =0.2 min: 1% B; t =3.0 min: 30% B; t =3.5 min: 50% B; t =5.0 min: 1% B。质谱监测器:SIM 模式(ESI 负离子),m/z 175.0;gain 1.0;裂片器70;干燥气体8.01/min,雾化器压力60 psi;干燥气体温度300℃;毛细管电压3 000 V。

第3课 Real-time multiplex SYBR Green Ⅰ-based PCR assay for simultaneous detection of *Salmonella* serovars and *Listeia moncytogenes*

1.原文

选自: Journal of Food Protection, 2003, 66(11):2141-2145

作者: Jothikumar N, Wang X W, Griffiths M W

Materials and Methods

Bacterial stains

Twenty-nine *Salmonella* strains; 18 strains of *L. moncytogenes*; 3 strains of *Campy-*

lobacter jejeuni; and strains of *E*. *coli*, *E*. *coli* O157:H7, *Yersinia enterocolitica*, *Shigella dysenteriae*, *Shewanella putrefaciens*, and *Proteus vularis* were obtained from the culture collection of the Canadian Research Institute for Food Safety, University of Guelph. All *Salmonella* and *L*. *monocytogenes* strains tested in the present study are listed in Table 1.

DNA extraction

A nonselective enrichment procedure followed by a rapid and simple DNA preparation for detection of *Salmonella* serovars and *L*. *moncytogenes* was followed. Crude DNA was extracted from all the strains grown overnight in nutrient broth (NB; BD Diagnostic Systems, Sparks, Md.) by the following procedure. Overnight broth culture (1 mL) was transferred to an Eppendorf tube and centrifuged at 12 000*g* for 2 min. Bacterial cells were resuspended in 100 μL of sterile deionized water, followed by boiling for 10 min. The heat-treated samples were chilled on ice for 2min and again centrifuged at 12 000*g* for 1 min. Clarified supernatants served as templates for PCR amplification.

Multiplex SYBR Green I PCR assay

The primer pair used in this study for detection of *Salmonella* was based on published sequence information. The primers SF (5′-CCTTT CTCCA TCGTC CTGA A-3′) and SR (5′-TGGTG TTATC TGCCT GACC-3′) were used to amplify an 85-bp sequence from the gene encoding a fimbrinlike protein (*fiml*) of *Salmonella enteritidis* (SA 942451). The primer pair LF: 5′-TCCGC AAAAG ATGAA GTTC-3′ (2539 to 22557) and LR: 5′-ACTCC TGGTG TTTCT CGATT-3′ (2636 to 2617) was designed in the present study to amplify a 98-bp sequence from the hemolysin gene (*hly*) of *L*. *moncytogenes* (GenBank accession no. M24199). The predicted length of each product was confirmed by agarose gel electrophoresis, and the T_m was used to identify specific products in subsequent analyses.

The PCR reaction was performed using a LightCycler (Roche Diagnostics, Laval, PQ, Canada) in glass capillary tubes containing 10 μL PCR reaction mixture consisting of 1 μL of FastStart SYBR Green enzyme and buffer (Roche Diagnostics). The reaction mixture also contained magnesium chloride (Mg^{2+}, 3 mmol/L; Roche), 0.25 μm of each primer, and 1 μL of DNA template. The conditions for the PCR reactions were an initial denaturation at 95 ℃ for 10 min, 45 cycles of 95 ℃ for 15 s, 55 ℃ for 5 s, and 72 ℃ for 10 s, followed by melting curve analysis. After PCR amplification, the PCR products were cooled to 65 ℃ and then slowly heated to 95 ℃ at a rate of 0.2 ℃/s. Fluorescence signals obtained were continuously monitored to confirm amplification specificity. In uniplex PCR, 0.5 μm of each primer was used.

2. 词或词组

agarose　*n*.　琼脂糖

Campylobacter jejeuni　空肠弯曲杆菌

clarified supernatant　上清液

denaturation　*n*.　使变性,钝化

fimbrinlike protein　类微丝蛋白

hemolysin gene 溶血素基因

L. moncytogenes 单核细胞增生李斯特菌

melting curve 溶解曲线

Proteus vularis 普通变形杆菌

primer *n.* 引物

PCR(polymerase chain reaction) 聚合酶链反应

real-time PCR 实时定量 PCR

strain *n.* 菌株

Salmonella 沙门菌

Shigella dysenteriae 痢疾志贺菌

Shewanella putrefaciens 腐败希瓦菌

Yersinia enterocolitica 结肠耶尔森菌

3.原文的参考译文

第3课 用于同步检测 *Salmonella* serovars 和 *Listeia moncytogenes* 的实时多元 SYBR Green Ⅰ型 PCR

材料和方法

细菌菌株

29株沙门菌,18株单核细胞增生李斯特菌,3株空肠弯曲杆菌,还有大肠杆菌,大肠杆菌O157:H7,结肠耶尔森菌,痢疾志贺菌,腐败希瓦菌,普通变形杆菌由加拿大圭尔夫大学的加拿大食品安全研究所菌种保藏中心提供。所有用于试验的沙门菌和单核细胞增生李斯特菌菌株都列于表1。

DNA 提取

非选择性扩大培养后,进行简单快速的DNA提取,用于沙门菌的检测,单核细胞增生李斯特菌也用相同方法操作。所有菌株在营养肉汤培养基中培养过夜,然后按下述方法提取DNA。取1 mL经过夜培养的菌液放入Eppendorf离心管中,在12 000*g*离心力下离心2 min。收集细菌细胞并悬浮于100 μL灭菌去离子水中,煮沸10 min后于冰上冷却2 min,再用12 000*g*离心力下离心1 min。取上清液作为PCR扩增的模板。

多重 SYBR Green I PCR 方法

用于检测沙门菌的引物选用已发表的序列。正向引物(5′-CCTTT CTCCA TCGTC CT-GA A-3′)和反向引物(5′-TGGTG TTATC TGCCT GACC-3′)设计源于肠炎沙门菌编码类微丝蛋白的基因,扩增产物为85 bp的序列。用于单核细胞增生李斯特菌检测的引物为自己设计,正向:5′-TCCGC AAAAG ATGAA GTTC-3′(2 539~22 557),反向:5′-ACTCC TGGTG TTTCT CGATT-3′(2 636~2 617)源于单核细胞增生李斯特菌溶血基因(GenBank编号:M 24199),扩增产物大小为98 bp。上述产物用琼脂糖凝胶电泳确认,在后续研究中用

退火温度(T_m)确定特定的产物。

PCR 反应在 LightCycler(Roche 公司,Laval, PQ,加拿大)中进行。反应管为玻璃毛细管,反应体积 10 μL,反应液中包含 1 μL FastStart SYBR Green Ⅰ酶和缓冲液(Roche 公司),还包括氯化镁(Mg^{2+},3 mmol/L;Roche 公司),0.25 μm 各引物和 1 μL DNA 模板。反应条件为:初始变性温度 95 ℃,维持 10 min。进行 45 个循环,每个循环为:95 ℃维持 15 s;55 ℃维持 5 s;72 ℃维持 10 s。之后进行溶解曲线分析。PCR 扩增结束后,扩增产物冷却到 65 ℃,然后缓慢升温到 95 ℃,升温速率 0.2 ℃/s。获得的荧光信号持续监测来确认扩增产物的特异性。在单一 PCR 中每种引物用量为 0.5 μm。

第4课　Changes in aroma characteristics of simulated beef flavour by soy protein isolate assessed by descriptive sensory analysis and gas chromatography

1.原文

选自: Food Research International,2007, 40: 1239-1248

作者: Moon S Y, Li-Chan E C Y

Materials and Methods

Sensory analysis

Descriptive sensory analysis was conducted by adapting the method of Zook and Pearce (1988) to obtain data describing the sensory attributes of SBF and SPI.

Panelist training

Ten subjects consisting of 8 women and 2 men with an interest in descriptive sensory evaluation were selected from students in the Food Science graduate program at the University of British Columbia and from food development staff at a food company producing soy-based meat alternative products for vegetarians and consumers preferring meatless products. The panelists received 8 h of training, consisting of four 2-h sessions conducted over 2 weeks.

***Training session* Ⅰ**

The objectives of the first training session were to discuss the aroma characteristics of the SBF and SPI and to propose descriptors for these attributes through open discussion. Two SBF samples, which were low SBF (150 mg in 5 g water) and high SBF (500 mg SBF in 5 g water), one SPI sample (500 mg SPI in 5 g water) and one mixture sample (500 mg SBF with 500 mg SPI in 5 g water) were prepared. Samples were provided in capped 15 mL vials covered with aluminum foil to minimize potential bias from sample appearance and held at 60 ℃ for 20 min before serving. The aroma descriptors for the SBF suggested by the panel at the first training session were (a) brothy/oxo/miso-like, (b) roasted/dry/cooked/barbecue, (c) beefy, (d) rare/raw/bloody/uncooked, (e) browned caramel/sweet/candy-like and (f) cardboardy/rancid/off-flavour/yeasty. The aroma descriptors suggested for the SPI were (a) soymilk-like, (b) cooked cereal, and (c) straw/hay-like.

2.词或词组

aluminum *n*. 铝

attribute *n*. 属性,特质

caramel *n*. 焦糖

cardboardy = cardboard flavour 纸板味

gas chromatography 气相色谱

miso *n*. 味增,日本豆酱

panelist *n*. 专题讨论小组参加者,感官鉴评成员

SBF(simulated beef flavour) 牛肉味香精

sensory analysis 感官分析

SPI(soy protein isolate) 大豆蛋白分离物

3.原文的参考译文

第4课 用感官分析和气相色谱评价大豆分离蛋白仿生牛肉的风味品质的变化

材料和方法

感官分析

感官分析采用 Zook 和 Pearce (1988)报道的方法,对 SBF 和 SPI 的感官特性进行描述。

品评人员培训

10 位品评人员包括对感官描述评估感兴趣的 8 位女性,2 位男性,他们主要来自于不列颠哥伦比亚大学食品专业的研究生、食品企业的产品研发人员。这些企业主要生产以大豆为基础的肉类替代产品,用来满足素食者或喜欢无肉产品的消费者的需求。品评人员要接受 8 h 的培训,4 次课程,每次 2 h,共耗时 2 周以上。

培训课程Ⅰ

第一次培训的目的是讨论 SBF 和 SPI 的风味特点,并通过开放式讨论来提出这些特征的描述词。准备 4 个样品:2 个 SBF 样品,1 个含量较低(150 mg/5 g 水),1 个含量较高(500 mg/5 g 水);1 个 SPI 样品(500 mg SPI/5 g 水)和 1 个混合样品(SBF 和 SPI 各 500 mg 于 5 g 水中)。准备好的样品放于 15 mL 的有盖小瓶中并用铝箔纸包裹,尽可能减少样品外观带来的偏差,然后于 60 ℃ 下放置 20 min 后用于品尝。经过第一次培训的小组成员建议对 SBF 风味的描述分为:(a)肉汤味/类似氧化味/味噌味,(b)烘烤味/肉干味/煮肉味/烧烤味,(c)牛肉味,(d)生味/血腥味,(e)焦糖味/甜味/糖味,(f)纸板味/酸败味/杂味/酵母味。对 SPI 的描述建议为:(a)豆浆味,(b)煮粥味,(c)稻草味/干草味。

第 5 课　Effect of processing on buckwheat phenolics and antioxidant activity

1.原文

选自：Food Chemistry，2006，99：388-393

作者：I'lkay Sensoy，Robert T. Rosen，Chi-Tang Ho，Mukund V. Karwe

Materials and Methods

Materials

Dark buckwheat flour：Dark buckwheat flour was purchased from Barry farm，Wapakoneta，Ohio，USA.

Whole buckwheat flour：Whole buckwheat grains（Barry farm，Wapakoneta，Ohio，USA）were milled using a Fitz Mill（Model D）or with a variable speed laboratory blender depending on the amount of flour needed.

White buckwheat flour：Hand separated groats from crushed whole buckwheat grain were milled[1] using a laboratory blender to make flour.

Corn meal：Corn meal was purchased from Iowa Corn Processors，Iowa，USA.

Whole wheat flour：Whole wheat flour，produced by Uhlmann Company（Kansas City，Missouri，USA）was purchased from a local store.

High amylopectin starch：High amylopectin starch（99% amylopectin）was purchased from National Starch and Chemical Company（Bridgewater，New Jersey. USA）.

Processes and analysis

All experiments were conducted in triplicate from the same batch of flour for all the subsequent analysis.

Roasting

Fifty grams of buckwheat flour was placed in a uniform thin layer（2-3 mm）in a 32 cm×23 cm cookie pan prior to roasting. It was roasted in a pre-heated oven（Fisher Scientific，Isotempé vacuum oven，model 282 A，USA）for 10 min at 200 ℃.

Extrusion

Since buckwheat starch has higher amylose content（46%）（Qian，Rayas-Duarte，& Grant，1998），it limits expansion during extrusion. Therefore，high amylopectin starch was mixed with dark buckwheat flour at 1∶1 ratio before the extrusion. The starch was added to increase the amylopectin to amylose ratio in the final mix to enhance expansion[2] during extrusion. The mixture was extruded with ZSK-30 twin-screw extruder（Coperion Corp.，Ramsey，New Jersey，USA）.

Flour was metered into the feed section of the extruder with a hopper feeder（K-Tron Corp.，Pitman，New Jersey，USA）. Water was injected into the feed section of the extruder using a triple action piston pump (US Electric Co., Milford CT). Throughput or the total mass flow rate (flour + water) was kept constant at 300 g/min during extrusion.

2. 词或词组

amylopectin *n*. 支链淀粉
amylose *n*. 直链淀粉
buckwheat *n*. 荞麦
extrusion *n*. 挤压
expansion *n*. 膨胀
hopper feeder 料斗送料器
groat *n*. 去壳谷粒
triple action piston pump 三联活塞泵
twin-screw extruder 双螺杆挤压机

3. 主要语法现象

[1] 现在分词短语作方式状语
[2] 动词不定式短语作目的状语

4. 原文的参考译文

第5课 加工对荞麦酚及其抗氧化活性的影响

材料和方法

原料

黑荞麦粉:黑荞麦粉购于美国俄亥俄州 Wapakoneta 的 Barry 农场。

荞麦全粉:荞麦谷粒购于美国俄亥俄州 Wapakoneta 的 Barry 农场,根据用量多少用 D 型 Fitz 粉碎机粉碎或用实验室捣碎器粉碎。

白荞麦粉:从捣碎的全麦粒中用手工挑出碎麦粒,然后用实验室捣碎器制粉。

玉米粉:购于美国艾奥瓦州,艾奥瓦玉米加工厂。

全麦粉:全麦粉购于当地市场,生产商为美国 Uhlmann 公司(堪萨斯市,密苏里州,美国)。

高级支链淀粉:高级支链淀粉Ⅰ(支链淀粉含量99%)购于美国新泽西州 Bridgewater 的国家淀粉和化学公司。

加工及分析

所有实验都做3次重复,包括同批次的面粉和所有后续分析。

烘烤

取50 g荞麦粉平铺于32 cm× 23 cm的烤盘中,厚度2~3 mm,然后置于预热的烤炉(Fisher Scientific, Isotempé 282A型真空烤炉,美国)中200 ℃烤制10 min。

挤压

由于荞麦粉直链淀粉含量较高，可达 46%，这样会限制其在挤压过程中的膨胀。所以在挤压前按 1∶1的比例将高级支链淀粉与黑荞麦粉混合，用以提高支链淀粉和直链淀粉的比例来增加挤压过程中的膨胀度。挤压在 ZSK-30 双螺杆挤压机(Coperion 公司，美国新泽西州 Ramsey)中进行。

将荞麦粉定量用料斗送料器加入喂料器(K-Tron 公司，美国新泽西州皮特曼市)。用三联活塞泵将水注入挤压机(US Electric Co.，米尔福德市)的喂料器中，在挤压过程中总物料(粉 + 水)速率维持在 300 g/min。

第 6 课　Changes in red wine soluble polysaccharide composition induced by malolactic fermentation

1. 原文

选自： Journal of Agricultural and Food Chemistry，2007，55:9592-9599

作者： Marguerite D L，Emmanuel G，Claire L M，Gérard C，Heyraud A，Lonvaud-Funel A

Soluble Polysaccharide (SP) Analysis.

The wine was centrifuged (10 000*g*，5 min，20 ℃) to eliminate insoluble matter. SP content was then analyzed without preliminary concentration，unless otherwise stated in the text. In this instance，15 mL of wine was put into Amicon Ultra tubes (Millipore) with a 5 ku membrane cutoff and centrifuged (5 000*g*，4 ℃) for as long as it took for the retentate to reduce to half the volume of the original sample. It was then analyzed like the other samples.

Three volumes of 96% ethanol containing 5% 1 mol/L HCl was added to the supernatant to precipitate the soluble polysaccharides. The tubes were left to stand for 24 h at 4 ℃. They were then centrifuged (18 000*g*，5 min，4 ℃)，and the pellet was washed with 80∶20 (*V*/*V*) ethanol/water，centrifuged again，dried for 20 min at 65 ℃，and dissolved in distilled water. The amount of total polysaccharides was determined by the phenol-sulfuric acid method，with glucose as the standard. The determination was done in triplicate. The result was considered as an estimated concentration because wine SPs contain glucose but also many other monosaccharides.

Samples for polysaccharide analysis were taken at the end of alcoholic fermentation just before inoculation with malolactic starters (initial estimated SP concentration) and at the stage of malic acid exhaustion (final estimated SP concentration). Delta SP is the difference between final and initial SP concentrations.

2. 词或词组

alcoholic fermentation　酒精发酵

centrifuge *n*. 离心
dissolve *v*. 使溶解,使溶化
distilled water 蒸馏水
ethanol *n*. 乙醇
insoluble *adj*. 不可溶的
malic acid 苹果酸
malolactic *adj*. 苹果乳酸的
monosaccharide *n*. 单糖
precipitate *v*. 使沉淀
retentate *n*. (在渗析过程中未能通过半透膜而被保留下的)保留物,滞留物,渗余物
soluble polysaccharide 可溶性多糖
starter *n*. 启动器,诱发剂
sulfuric acid 硫酸

3.原文的参考译文

第6课 苹果酸发酵引起的红葡萄酒中可溶性多糖的变化

可溶性多糖的分析

将葡萄酒离心(10 000*g*, 5 min, 20 ℃)去除不溶性的物质,然后分析可溶性多糖的含量,除非特殊说明一般不需要浓缩。在本次试验中,取 15 mL 葡萄酒移入 Amicon Ultra 超滤管(Millipore)配有 5 ku 的滤膜,并且长时间离心(5 000*g*, 4 ℃),充分提出滞留物,直至为原体积的一半,然后像其他样品一样进行测定。

将 3 倍体积的含 5% 1 mol/L HCl 的 96% 乙醇溶液加入到上清液中沉淀可溶性多糖。将试管在 4 ℃下放置 24 h。然后离心(18 000*g*, 5 min, 4 ℃),沉淀用 80:20(体积分数)的乙醇水溶液洗涤,离心,65 ℃干燥 20 min 后,溶于蒸馏水中。用苯酚-硫酸法测定总多糖的含量,用葡萄糖作标准,测定重复进行 3 次。测定结果为大概的浓度,因为在葡萄酒可溶性多糖中除了葡萄糖外,还有许多种其他单糖。

多糖测定的取样包括:①乙醇发酵结束后,即在接种乳酸菌进行苹果酸-乳酸发酵之前取样,用于测定起始可溶性多糖浓度;②在苹果酸-乳酸发酵结束时取样,用于测定最终可溶性多糖的浓度。Delta 可溶性多糖为最终和起始可溶性多糖之差。

作 业 汉译英

1.本文研究了一种新的 DNA 提取和纯化的方法,并结合多重实时 PCR 同时检测生肉中沙门菌和单核细胞增生李斯特菌。引物设计选用沙门菌的 *inv*A 基因和单核细胞增生李斯特

菌的 *hly*A 基因。扩增产物检测用 SYBR Green 荧光和溶解温度。结果显示，此方法在 10 h 内的检测量为每克样品香肠中含有 2 个单核细胞增生李斯特菌细胞和 4 个沙门菌细胞，其中包括 6～8 h 的富聚阶段。

2.本文研究了加工对荞麦功能成分的影响。在烘烤或挤压前后分别萃取荞麦面粉得萃取物。用 Folin-Ciocalteu 测定总酚，发现加工不会改变荞麦粉中的总酚含量。200 ℃烘烤 10 min 后，黑荞麦粉中非极性成分和极性成分都有所增加，而挤压后只有极性成分增加。抗氧化活性(DPPH)结果显示：200 ℃烘烤 10 min 会轻微的降低其抗氧化活性，而挤压(170 ℃)不会对其抗氧化活性产生任何影响。所以认为可以通过优化工艺条件来保持荞麦产品中有益于身体健康的成分。

3.将 2.5%、5%、7.5%和 10%的可溶性和不溶性非淀粉多糖分别添加在意大利面条的面团中，采用一系列分析方法分析其烹调和组织结构特性。发现在通常情况下非淀粉多糖会增加烹调损失，降低面中蛋白质和淀粉含量。这种影响取决于非淀粉多糖的添加量和多糖的形式(可溶或不溶)。虽然一些胶类非淀粉多糖会提高面条的硬度，但一般情况下面条的硬度下降与非淀粉多糖的添加量有关。面条的黏性、韧性及弹性也会受到影响。研究结果表明在意大利面中强化非淀粉多糖一定要谨慎选择添加量，确保最佳的组织结构和烹调特性。

第 7 章

论文的结果与讨论

Results and Discussion of Article

第 1 课　The effect of steeping time on the final malt quality of buckwheat

1.原文

选自： J. Inst. Brew, 2005, 111(3): 275-281

作者： Wijngaard H H, Ulmer H M, Neumann M, Arendt E K

Total Nitrogen (TN). In barley, besides protein many other nitrogenous compounds are present, such as amino acids and nucleic acids. Apart from substances leached from the grain during steeping, there is no loss of nitrogen from the barley grain during malting and the respiratory loss of dry matter tends to raise the TN of the grain. However, nitrogenous compounds move into the rootlets during malting and the removal of rootlets after malting leads to a significant loss of TN in the finished malt. TN of buckwheat malt decreased with increasing SMC (Table Ⅱ). This can be explained by the fact[1] that at higher SMC's more nitrogenous compounds were used for rootlet growth. It is assumed[2] that a similar mechanism occurs in buckwheat as in barley and nitrogenous compounds are moving into the rootlets during germination. Since the rootlets of the buckwheat malts were removed, more TN was removed with the longer rootlets.

Viscosity. During barley malt brewing 1, 3-1, 4-mixed β-glucans are particularly important because these β-glucans can: ⅰ) reduce extract yields in the brewhouse; ⅱ) participate in haze formation in wort and beer and ⅲ) cause high wort and beer apparent viscosities. These high apparent viscosities can lead to lautering and beer filtration problems. The 1,3-1,4-mixed β-glucan is mainly found in the endosperm cell walls[3] that surround barley starch granules. Like barley starch granules, buckwheat starch granules are located in cells in the endosperm and are surrounded by relatively thin cell walls. In contrast to barley, these cell walls do not contain 1,3-1,4-mixed β-glucan. Nevertheless, the viscosity of buckwheat congress worts was significantly higher than congress worts[4] produced with barley malt. This could indicate the presence of a polysaccharide other than 1,3-1,4-mixed β-glucan in buckwheat wort. Asano et al have identified a polysaccharide in the buckwheat grain[5] that consisted of xylose, mannose, galactose and glucuronic acid. Although the proportion of soluble fibre was relatively small (2.9%), these soluble fibre compounds cause high viscosities and are therefore very significant for the brewing process. Table Ⅱ shows the viscosity of congress worts[6] produced from buckwheat malts[7] steeped at different moisture contents. It is clearly seen[8] that with increased SMC, wort viscosity decreased. This could be explained by the fact[9] that polysaccharides might be degraded to a higher extent in worts[10] derived from buckwheat malts with higher SMC's. When malting barley, a reduction in β-glucan content of worts was reported when malting with higher SMC's The authors found a significant relation between *β*-glucan contents and viscosities of worts.

Table Ⅱ Average values of trial Ⅰ and Ⅱ of characteristics of B35, B40, B45 and control barley malt

Parameter	B35	B40	B45	Barley malt
Rootlet length after 60 h germination(mm)	13.4	17.2	29.2	n.a.
Malting loss(%)	7.43	7.89	10.74	n.a.
Moisture percentage kined and cleaned malt(%)	5.53	5.48	7.51	3.59
TN(%)	2.38	2.32	2.31	1.59
Wort viscosity(mPa · s)	2.34	2.22	2.07	1.61
Friability(%)	96.95	95.42	91.22	89.36
Alpha-amylase activity(international units g^{-1} wet wt.)	204.26	204.26	237.79	582.72
Filterability	poor	poor	poor	poor
Extract d.w.(%)	63.77	65.57	63.68	81.85
Fermentability(%)	56.11	55.99	53.13	82.20
TSN(%)	0.55	0.55	0.56	0.50
Kolbach Index(%)	23.09	23.91	24.18	31.24
FAN($mg \cdot L^{-1}$)	100.94	106.89	99.95	127.31

n.a. = not applicable

2.词或词组

barley *n*. 大麦

brewhouse *n*. 啤酒厂

buckwheat *n*. 荞麦

endosperm *n*. 胚乳

germination *n*. 发芽

β-glucan *n*. *β*-葡聚糖

granule *n*. 淀粉颗粒

malting *n*. 麦粒发芽,制成麦芽糖,麦芽作坊;麦芽制造(法)

polysaccharide *n*. 多糖

rootlet *n*. 小根,根芽

steeping *v*. 浸麦; *adj*. 浸渍的

total nitrogen (TN) 总氮

the respiratory loss 呼吸损失

viscosity *n*. 黏度

wort *n*. 麦芽汁

3.主要语法现象

[1][3][5] 定语从句

[2] 宾语从句

[4][6][7][9][10] 过去分词作定语，放在被修饰词的后面

[8] 形式主语从句

4.原文的参考译文

第 1 课　浸麦时间对荞麦麦芽品质的影响

总氮。在大麦中，除了蛋白质外，还存在其他的含氮物质，如氨基酸和核酸。除了在浸麦时滤除了谷物中的杂质外，在制麦期间啤酒大麦籽粒没有氮的损失，干物质的呼吸损失使得籽粒的总氮含量有所上升。然而，含氮物质在制麦过程中转移到根芽中，在制麦结束后去除根芽导致在成品麦芽中总氮显著地减少(表 2)。荞麦麦芽的总氮含量随着浸麦度的增加而降低，这是由于在高浸麦度时，需要更多的含氮物质用于根芽的生长。推测如同啤酒大麦一样，荞麦在发芽过程中含氮物质转移到根芽中。去除了荞麦麦芽的根芽导致总氮损失，根芽越长，总氮损失得越多。

黏度。在啤酒大麦麦芽酿造过程中，1,3-1,4-复合 β-葡聚糖尤为重要，因为这些 β-葡聚糖能够：①降低啤酒厂的浸出物产量；②参与麦芽汁形成和啤酒混浊的产生；③导致麦芽汁和啤酒的表观黏度增加，高的表观黏度造成啤酒的澄清和过滤问题。1,3-1,4-复合 β-葡聚糖主要存在于包围大麦淀粉颗粒的胚乳细胞壁中。与大麦淀粉颗粒一样，荞麦淀粉颗粒位于胚乳中的细胞中，被相对较薄的细胞壁包围。相比于啤酒大麦，这一类细胞壁并没有包含 1,3-1,4-复合 β-葡聚糖。可是荞麦麦芽汁的黏度比啤酒大麦麦芽汁要高出很多，这可能意味着在荞麦麦芽汁中存在其他的多糖而不是 1,3-1,4-复合 β-葡聚糖。Asano 等已经鉴定出荞麦籽粒中的多糖组分有木糖、甘露糖、半乳糖和葡萄糖醛酸等。尽管荞麦中的可溶性纤维含量较低(2.9%)，但这些可溶性纤维可以造成麦芽的黏度较高，这对啤酒厂非常重要。表 2 是在不同水分含量浸麦时荞麦麦芽的麦芽汁黏度结果，从中可以看出随着浸麦度的增加，麦芽汁黏度降低，说明较高浸麦度的荞麦麦芽中的多糖被降解。对于啤酒大麦，已经被报道在较高浸麦度时，β-葡聚糖含量较低。作者发现 β-葡聚糖含量和麦芽汁黏度有显著的相关性。

表 2　B35,B40,B45 和对照大麦麦芽品质的试验Ⅰ和试验Ⅱ的平均值

参　数	B35	B40	B45	大麦麦芽
发芽 60 h 后根芽长度/mm	13.4	17.2	29.2	n.a.
制麦损失/%	7.43	7.89	10.74	n.a.
成品麦芽水分含量/%	5.53	5.48	7.51	3.59
总氮/%	2.38	2.32	2.31	1.59
麦芽汁黏度/(mPa·s)	2.34	2.22	2.07	1.61
脆度/%	96.95	95.42	91.22	89.36
α-淀粉酶活性/(U/g 湿重)	204.26	204.26	237.79	582.72
过滤性	较差	较差	较差	较差
无水浸出物/%	63.77	65.57	63.68	81.85

续表 2

参　数	B35	B40	B45	大麦麦芽
发酵度/%	56.11	55.99	53.13	82.20
可溶性氮/%	0.55	0.55	0.56	0.50
库尔巴哈值/%	23.09	23.91	24.18	31.24
FAN/(mg/L)	100.94	106.89	99.95	127.31

注：n.a. = 未测定

第 2 课　Volatile and non-volatile chemical composition of the white guava fruit (*Psidium guajava*) at different stages of maturity

1. 原文

选自： Food Chemistry，2007，100(1)：15-21

作者： Soares F D，Pereira T，Maio Marques M O，Monterro A R

3.1　Non-volatiles analyses

Table 1 shows the results of color，pH，titratable acidity，total soluble solids，sugars and vitamin C in white guava fruits at different stages of maturity. The results are the average of three replications.

3.1.1　Fruit color

The color of white guava fruits was expressed by Hunter parameters (L，a，b，c and h^o). When data a and b were converted to chroma (c) the values became positive. The loss of green color and increase of yellow color is evidenced by the increase of values chroma from 24.3 to 33.10. L parameters increased from 53.83 to 71.87. The hue angle (h^o) values obtained were：113.73，95.91 and 82.66 for immature，intermediate and mature fruits，respectively. Statistic analysis of the results showed significant effect ($P \leqslant 0.05$) in the three different stages. These results are according to Mattiuz and Durigan (2001a)[1] who studied guava fruits of the varieties "Pedro Sato" and "Paluma"[2] stored at room temperature. Pereira (2003) and Brecht (1980) related that the color of fruits is associated with synthesis and degradation of pigments. Pereira (2003) showed[3] that for white guava fruits the green color decreased along with the decrease of chlorophyll level and yellow color increased with the increase of carotenoid level.

3.1.2　pH and Titratable acidity

The pH increased slowly during the different maturity stages whereas titratable acidity increased in the immature and intermediary stage of maturation and decreased in the maturity stage. Increase in the pH and titratable acidity show the formation of organic acids during maturation.

These increases on both parameters are associated with high concentration of undissociated organic acids, stored in the vacuole and the fruits use these acids as respiratory substrate (Medlicott & Jeger, 1987). Bashir, Abu-Goukh, and Abu-Bakr (2003) reported similar results to pink and white guava pulp. Titratable acidity increased from 0.15% to 0.19% of citric acid up to the climacteric peak and declined thereafter from 0.19% to 0.154% citric acid for white guava.

3.1.3 Total soluble solids

Total soluble solids increased during ripening (7.40-8.6 °Brix), but on the maturity stage decrease to 8.4 °Brix. This probably happened because of high consumption of sugars due to respiration rate (Sharaf & El-Saadany, 1996; Singh, Singh, and Chauan, 1981). Rodrigues, Agarwal, and Saha (1971) reported similar behavior for guava cultivar Sefeda. The authors showed a gradual increase of the content of soluble solids of guava fruit with maturation, except during the end of the growth period. According to El Bulk, Babiker, and El Tinay (1997) for different cultivars of guava total soluble solids content gradually increased with fruit development in all cultivars. The results obtained were: between 7.70, 6.20, 6.6 and 9.70 °B for Shambati, Pakistani, Shendi and Ganib cultivars, respectively, after first 15 days post harvest and 13.2, 11.1, 12.2 and 12.5 °B when the fruits were 126 days. The authors observed an increase in total soluble solids after 106 days.

2.词或词组

acidity *n.* 酸度
consumption *n.* 消费,消费量
cultivar *n.* 栽培品种
guava *n.* 番石榴
maturity *n.* 成熟
organic acid 有机酸
postharvest *adj.* 采后的
respiration rate 呼吸速率
titratable *adj.* 可滴定的
total soluble solids 总可溶性固形物
vacuole *n.* 液泡
volatile *adj.* 挥发性的; *n.* 可挥发物

3.主要语法现象

[1] 由 who 引导的限制性定语从句
[2] 过去分词作定语,放在被修饰词的后面
[3] 由 that 引导的宾语从句

4.原文的参考译文

第2课　不同成熟阶段的白番石榴中的挥发性和非挥发性化学成分

3.1　非挥发性组分的分析

表1是不同成熟期的白番石榴的色泽、pH、可滴定酸、总可溶性固形物、糖和维生素C含量。结果用3次重复测定的平均值表示。

3.1.1　果实的色泽

白番石榴的色泽用Hunter值表示(L，a，b，c和h^{o})。当数值a和数值b被转化为色度值c时，数值成为正值。色度值从24.3增加到33.10，可以说明绿色色调的降低和黄色色调的增加。L值从53.83增加到71.87。色彩角值(h°)在未成熟期、中等成熟期和成熟期分别为113.73、95.91和82.66。数据统计结果表明在3个不同的阶段有显著的差异性。这些结果与Mattiuz和Durigan对研究番石榴品种"Pedro Sato"和"Paluma"的结果是一致的。Pereira(2003)和Brecht(1980)论述了果实的色泽与色素的合成和降解是相关联的。Pereira(2003)证实了随着叶绿素水平的降低白番石榴果实的绿色降低，而随着胡萝卜色素水平的增加，黄色也在增加。

3.1.2　pH和可滴定酸

在不同的成熟阶段，pH缓慢增加，然而可滴定酸在未成熟期和成熟的中间阶段增加，在成熟期降低。pH和可滴定酸增加说明在成熟过程中有机酸的形成。这2个指标的上升与未解离的有机酸浓度高有关，有机酸贮存于液泡中，果实利用这些酸作为呼吸基质(Medlicott和Jeger，1987)。Bashir，Abu-Goukh和Abu-Bakr(2003)研究也表明粉色和白色番石榴果浆有类似结果。白色番石榴的可滴定酸从0.15%增加至0.19%(以柠檬酸计)，达到峰值后又从0.19%降至0.154%。

3.1.3　总可溶性固形物

果实在成熟过程中(7.40～8.6 °Brix)，总可溶性固形物不断增加，但到了成熟期就降为8.4 °Brix，可能是因为呼吸强度增加了糖的消耗(Sharaf和El-Saadany，1996；Singh，Singh和Chauan，1981)。这与Rodrigues，Agarwal和Saha(1971)报道的番石榴品种Segeda结果类似。研究者们报道：除在果实生长末期外，可溶性固形物含量随着果实的不断成熟而逐渐增加。El Bulk，Babiker和El Tinay(1997)等研究了不同的番石榴品种可溶性固形物的含量，结果表明所有的番石榴品种在果实成熟过程中总可溶性固形物含量逐步增加，品种为Shambati、Pakistani、Shendi和Ganib的番石榴采后15 d的总可溶性固形物含量分别为7.70、6.20、6.6和9.70 °Brix，在106 d后总可溶性固形物含量开始升高，在126 d后的总可溶性固形物分别为13.2、11.1、12.2和12.5 °Brix。

第 3 课　Sugars, organic acids, phenolic composition and antioxidant activity of sweet cherry (*Prunus avium* L.)

1.原文

选自： Food Chemistry, 2008, 107: 185-192

作者： Valentina Usenik, Jerneja Fabčič, Franci Štampar

3.2　Sugars

Glucose, fructose, sorbitol and sucrose contents of the sweet cherry fruits are shown in Table 2. Generally, glucose was found to have the highest content, followed by fructose, sorbitol and sucrose,[1] confirming the results of Serrano et al. (2005) and Usenik, Stampar, Sturm, and Fajt (2005). Cultivar Early Van Compact had the highest and Sylvia the lowest glucose content. The content of fructose varied from 47.6 (Sylvia) to 102 g/kg FW (Lala Star). The content of sorbitol varied from 4.45 (Ferprime) to 26.7 g/kg FW (Early Van Compact). Cultivar Lala Star had the highest and Sylvia the lowest sucrose content. The highest sum of sugars was found in Lala Star and Early Van Compact and the lowest in Sylvia.

3.3　Organic acids

The results show variations between cultivars (Table 3). Malic, citric, shikimic and fumaric acids were all detected in sweet cherry cultivars. The predominant organic acid in sweet cherry was malic acid,[2] which is in agreement with Serrano et al. (2005). The content differed among cultivars: malic acid 3.53-8.12 g/kg FW, citric acid 0.11-0.54 g/kg FW, shikimic acid 6.56-26.7 mg/kg FW and fumaric acid 0.97-7.56 mg/kg FW. Cultivar Fercer had the highest content of malic, citric and shikimic acid and a high content of fumaric acid. The lowest content of malic acid was measured in Lapins and the lowest content of citric acid in Vesseaux. Similar content of acids for cultivars Burlat, Vigred and Lapins were measured in the study of Sturm and Stampar (1998).

3.4　Phenolic compounds

The content of phenolic compounds detected in sweet cherry fruits is shown in Table 4. Neochlorogenic acid, p-coumaroylquinic acid and chlorogenic acid (hydroxycinnamic acids), epicatechin (flavan-3-ol) and rutin (flavonol) were analysed in the study. Generally, the same phenolic compounds were present in each cultivar, but there were differences in relative levels. Neochlorogenic acid was the major hydroxycinnamic acid derivative ranging from 4.74 to 11.9 mg/100 g FW, followed by p-coumaroylquinic acid (0.77-7.20 mg/100 g FW) and chlorogenic acid (0.60-2.61 mg/100 g FW). The contents of neochlorogenic acid, p-coumaroylquinic acid and chlorogenic acid are similar to those of Kim et al. (2005). Our results for phenolic acids content, are low compared to those of

Gonc-alves et al. (2004a), but the proportions are similar. Neochlorogenic acid comprised 24%-65%, p-coumaroylquinic acid 5%-31% and chlorogenic acid 3%-15% of the phenolics. Cultivar Early Van Compact was found to have a significantly higher content of neochlorogenic acid,[3] compared to Badascony, Burlat, Fercer, Noire de Meched, and Vigred. Cultivars Early Van Compact, Fernier, Ferrador and Lapins had a significantly lower content of p-coumaroylquinic acid, compared to cultivars Burlat, Ferprime, Noire de Meched and Sylvia. Cultivars Fernier, Lala Star, Early Van Compact and Vesseaux had a significantly higher content of chlorogenic acid, compared to cultivars Badascony, Burlat, Fercer, Ferprime, Ferrador, Noire de Meched, Sylvia and Vigred. The highest contents of epicatechin (from 0.43 to 4.51 mg/100 g FW) were found in cultivars with early ripening time (Ferprime, Burlat, Vigred).

2.词或词组

antioxidant activity　抗氧化活性

be correlated with　与……相关联,把……与……联系起来

confirm　*v.*　确定,确认

in agreement with　符合……,一致

sweet cherry　甜樱桃

total phenolic content　总多酚含量

3.主要语法现象

[1] 现在分词短语作独立结构

[2] which 引导的非限制性定语从句

[3] 过去分词短语作状语

4.原文的参考译文

第3课　甜樱桃(*Prunus avium* L.)中糖、有机酸、酚类物质和抗氧化活性

3.2　糖

甜樱桃中的葡萄糖、果糖、山梨糖醇和蔗糖的含量见表2。一般来说,含量最高的糖是葡萄糖,其次是果糖、山梨糖醇和蔗糖,这与 Serrano 等的研究结果一致。品种 Early Van Compact 含有最高的葡萄糖含量,而品种 Sylvia 的葡萄糖含量最低。果糖的含量变化范围为47.6～102 g/kg 鲜重(Lala Star)。山梨糖醇的含量变化范围为4.45～26.7 g/kg 鲜重(Early Van Compact)。品种 Lala Star 具有最高的蔗糖含量,品种 Sylvia 具有最低的蔗糖含量。含糖量最高的品种是 Lala Star 和 Early Van Compact,最低的是 Sylvia。

3.3　有机酸

品种间的有机酸含量变化见表3。

在甜樱桃品种中主要检测了苹果酸、柠檬酸、莽草酸、富马酸的含量。甜樱桃中含有的有

机酸主要是苹果酸，这与Serrano等(2005)报告的结果一致。在品种间不同的有机酸含量变化为：苹果酸为3.53～8.12 g/kg鲜重，柠檬酸为0.11～0.54 g/kg鲜重，莽草酸为6.56～26.7 mg/kg鲜重，富马酸为0.97～7.56 mg/kg鲜重。品种Fercer中苹果酸、柠檬酸和莽草酸含量最高，富马酸含量较高。品种Lapins中苹果酸含量最低，品种Vesseaus中柠檬酸含量最低。根据Sturm和Stampar的研究，品种Vigred、Lapins和Burlat中所含有的酸类成分相似(1998)。

3.4　酚类成分

甜樱桃中酚类成分的检测结果见表4，主要分析了新绿原酸、对香豆酰奎宁酸、绿原酸、表儿茶酸和芦丁含量。通常情况下，每个品种中均含有相同的酚类物质成分，但含量有所不同。新绿原酸是主要的羟基苯乙烯酸，含量的变化范围为4.74～11.9 mg/100 g鲜重，其次为对香豆酰奎宁酸(0.77～7.20 mg/100 g鲜重)，绿原酸(0.60～2.61 mg/100 g鲜重)。新绿原酸、对香豆酰奎宁酸、绿原酸的含量测定结果与Kim等(2005)的相似。就酚类物质含量而言，我们的测定结果低于Concalves等(2004a)的分析结果，但各成分间的比例是相似的。酚类成分中包含24%～65%的新绿原酸，5%～31%的对香豆酰奎宁酸，3%～15%的绿原酸。分析结果发现，相比较于品种Badascony、Burlat、Fercer、Noire de Meched和Vigred，品种Early Van Compact中含有相当高的新绿原酸。同品种Burlat、Ferprime、Noire de Meched和Sylvia相比，品种Early Van Compact、Fercer、Ferrador和Lapins中对香豆酰奎宁酸含量很低。与品种Badascony、Burlat、Fercer、Ferprime、Ferrador、Noire de Meched相比较，品种Fernier、Lala Star、Early Van Compact和Vesseaux中含有较高的绿原酸。在早熟品种(Ferprime、Burlat、Vigred)中表儿茶酸含量(0.43～4.51 mg/100 g鲜重)最高。

第4课　Evaluation of processing qualities of tomato juice induced by thermal and pressure processing

1.原文

选自：LWT-Food Science and Technology，2008，41(3)：450-459

作者：Hsu K C

3.1　Color

Effects of thermal and pressure processing on the red-yellow ratio,[1] indicating the redness of tomato juices, are shown in Fig. 1. An a/b ratio of 1.90 or greater represents a first quality product in terms of color and an a/b ratio of less than 1.80 means[2] that the tomato products may be unacceptable for inclusion in products[3] where a bright red color is desired. The a/b value of the control was 3.62 and appreciated more than that of hot-break tomato juice of 3.30 ($P=0.05$) and coldbreak tomato juice of 3.54. The result in a low a/b value represented an orange to brown color due to the breakdown of lycopene and formation of Maillard reaction products by the intensive heat treatment. The a/b values of the pressure-treated samples were significantly greater than that of control or thermally treated samples. And a/b values of tomato juice increased up to 3.84 with pressure levels elevated to 500 MPa. Results suggest[4] that an increase in the red color (a value; data not shown) of high pressure treated tomato juice compared to thermal treatments, attributed to the better homogenization and brightening of the red color. However, contradictory results can be found in literature about the effect of high pressure on other fruits and vegetables.[5] Combined high pressure and thermal treatments at 300-700 MPa/65 ℃ for 60 min did not significantly change the $L^* a^* / b^*$ parameter of tomato puree, however, 600-700 MPa significantly increased the same parameter of strawberry juice at pH 5.0 for 8.8% (Rodrigo et al., 2007). A high retention of red color was observed when high pressure treating strawberries (Matser & Bartels, 1999). Also, high pressure has been found to induce discoloration in mushrooms and onions because of the activity of the enzyme polyphenoloxidase, responsible for browning (Butz, Koller, & Tauscher, 1994). Those discrepancies might be due to the different color contributors, color degradations and various processing conditions, including pH, time and temperature (Rodrigo et al., 2007).

3.2 Carotenoids and lycopene

The total carotenoids and lycopene contents of control are 212.8 and 145.6 mg/g, respectively (data not shown),[6] which are higher than those of tomato (Tau-Tai Lan T93) juices by about 20% probably due to different cultivars (Lin & Chen, 2003). After the cold and hot break, both total carotenoids and lycopene contents of tomato juices slightly but insignificantly decreased about 1% ($P = 0.05$) (Fig. 2). However, the positive effect of temperature on the extractability of lycopene is described in the literature, this effect being time-depending (Porrini, Riso, & Testolin, 1998). Increases in lycopene concentration in tomato puree have been shown at 90 ℃/110 min and 110 ℃/1.1 min but not at 120 ℃/0.1 min (Anese, Falcone, Fogliano, Nicoli, & Massini, 2002). The results showed[7] that cold and hot break[8] applied to tomato juice in this study did not induce lycopene extractability or degradation probably owing to insufficient temperature and time. Further, the color degradation[9] occurring by both thermal treatments in this study was caused by nonenzymatic reactions (Fig. 1).

2. 词或词组

carotenoid *n.* 类胡萝卜素
cold-break 冷破碎
breakdown *n.* 降解
homogenization *n.* 均质化,均匀化,同质化
hot-break 热破碎
lycopene *n.* 番茄红素
nonenzymatic reaction 非酶反应
thermal *adj.* 热的,热量的
tomato juice 番茄汁

tomato puree　番茄酱

3. 主要语法现象

[1] 现在分词作状语

[2][4][7] 宾语从句

[3] 定语从句

[5] 过去分词短语作主语

[6] 非限制性定语从句

[8] 过去分词作宾语

[9] 现在分词短语作定语

4. 原文的参考译文

第 4 课　热加工和高压加工对番茄汁加工品质的影响

3.1　色泽

热处理和压力处理对表示番茄汁红色的红-黄值的影响结果见图 1。a/b 值为 1.90 或高于 1.90 表示的色泽是番茄产品的首要品质，若 a/b 值低于 1.80 则认为产品不被消费者所接受，番茄产品的亮红色是理想的颜色。对照样品的 a/b 值为 3.62，比热破碎番茄汁 3.30（$P=0.05$）和冷破碎番茄汁 3.54 要高一点。低的 a/b 值说明色泽由橙色变为棕色，这是由于在强热处理下番茄红素降解，并且形成了美拉德反应产物。经压力处理下的样品其 a/b 值显著高于对照样品或热处理样品。随着压力水平上升到 500 MPa，番茄汁的 a/b 值增加到 3.84。试验结果表明，相对于热处理，高压处理的番茄汁其红色有所增加，有助于更好地均质和增加红色的亮度。不过，有文献报道了关于高压对其他水果和蔬菜的影响与本研究的结果完全相反。在 300～700 MPa 压力和 65 ℃ 的温度处理 60 min 并没有显著地改变番茄酱的色度，而在 600～700 MPa 压力下处理 pH 为 5.0 的草莓汁时可使其色度值增加 8.8%。当高压处理草莓时，红色的保持性更好。同时，研究还发现多酚氧化酶、高压等因素导致了蘑菇和洋葱的变色。这些差异可能由不同的色素来源、色素降解以及不同的加工条件，包括 pH、时间和温度所引起的。

3.2　类胡萝卜素和番茄红素

对照样品的总类胡萝卜素和番茄红素的含量为 212.8 mg/g 和 145.6 mg/g，比起那些来自不同品种的番茄汁要高出 20%。经过冷破碎和热破碎 2 种处理后的番茄汁，其总类胡萝卜素和番茄红素含量显著地降低了 1%（$P<0.05$）（图 2）。然而，据文献报道，温度对于番茄红素的萃取有促进作用，且这种作用取决于萃取时间，将番茄酱在 90 ℃ 作用 110 min，110 ℃ 下作用 1.1 min，番茄酱中的番茄红素含量有所增加，但 120 ℃、0.1 min 的处理组没有。研究结果表明，本研究中采用的冷破碎和热破碎方式加工的番茄汁并没有影响番茄红素的萃取或降解，可能是由于温度和时间不够充分。此外，在 2 种热处理中发生的色素降解是由非酶褐变引起的（图 1）。

第5课 Changes of pesticide residues in apples during cold storage

1.原文

选自：Food Control，2008，19(3)：247-256

作者：Ticha J，Hajslova J，Jech M，Honzicek J，Lacina O，Kohoutkova J，Kocoure K V，Lansky M，Kloutvorova J，Falta V

Post-harvest diseases can be a limiting factor for the long-term storage of apples. As mentioned in Introduction，orchard practices such as sanitation and fungicide application as well as a strategy of insects control can have a great impact on the types and amount of decay potentially occurring during cold post-harvest storage.[1] Regarding crop storage life-time，fungicides applied near harvest time may provide some control of damage-causing pathogens originated both from the latest fungal infection of fruit in the orchard and those developed by fungal infection of wounds（punctures，bruises etc.）caused by harvest and post-harvest handling practices. However，it should be emphasized that besides of benefits obtained by chemical crop protection，also health hazards associated with pesticides use have to be taken into consideration. To meet both consumers and toxicologists concerns，residues potentionally[2] occurring in food supply have to be controlled. While in our previous study（Ticha et al.，2006）the dynamics of pesticide residues in apples within the pre-harvest period was investigated，and treatment regimes leaving minimum residues in fruit intended for direct consumption and/or baby food production were searched，in the current study we focused on the fate of residues during the post-harvest period，under conditions of cold storage. The overview of detected pesticide residues used for orchard treatment in field experiments FT1，FT2 and FT3 in apples at the time of harvest is shown in Fig. 1.（In field experiment FT4，none of pesticides used for apple trees treatment left detectable residues at the harvest time.）Of 21 active ingredients of pesticide preparations used for apple trees protection（see Table 1 for detailed information），only six fungicides and one insecticide were found in apples at the beginning of storage period. Residues of acetamiprid，chlorpyrifos-methyl，difenoconazole，diflubenzuron，dithianon，EBDCs（represented by mancozeb and thiram in this study），fenoxycarb，kresoxim-methyl，teflubenzuron，thiacloprid，triazamate，trifloxystrobin and triflumuron dropped below detection limits of analytical methods employed in this study（see Table 5 for overview of LOQs of respective analytes）. Further decrease of residues occurred during cold storage，see Fig. 2；dodine was only one of those[3] detected fungicides found after 5 months in one of experiments. High persistency was also documented for organophosphorus pesticide phosalone (see Fig. 2). In both cases, pesticide residues were below 0.01 mg/kg that is maximum residue limit required by baby food producers for raw material to be processed. In the following paragraphs (Figs. 1-6), more detailed information on pesticides[4] we monitored during post-harvest period is provided. Worth to notice, that compounds representing various chemical classes were selected. They are commonly used in conventional apple orchards and are often detected in matured apples within surveillance programs.

2. 词或词组

acetamiprid　*n.*　啶虫脒(杀虫剂)

chlorpyrifos-methyl　*n.*　甲基氯吡磷(杀虫剂)

decay　*v/n.*　腐朽,腐烂,衰退,衰减

difenoconazole　*n.*　苯醚甲环唑

diflubenzuron　*n.*　除虫脲

dithianon　*n.*　二氰蒽醌

dodine　*n.*　多果定(杀虫剂)

EBDCs　二硫代氨基甲酸(盐)类杀菌剂

fenoxycarb　*n.*　苯氧威,又名双氧威、苯醚威

fungicide　*n.*　杀菌剂

kresoxim-methyl　醚菌酯,一种高效、广谱、新型杀菌剂

long-term　长期的

mancozeb　*n.*　代森锰锌

orchard　*n.*　果园

organophosphorus pesticide phosalone　有机磷农药

pesticide　*n.*　杀虫剂

post-harvest　*n.*　采后

sanitation　*n.*　卫生,卫生设施

thiram　*n.*　福美双(一种广谱保护性的福美系杀菌剂)

teflubenzuron　*n.*　伏虫隆

thiacloprid　*n.*　噻虫啉

triazamate　*n.*　唑蚜威

trifloxystrobin　*n.*　戊菌酯

triflumuron　*n.*　杀虫脲,氟幼灵

3. 主要语法现象

[1] 现在分词短语作状语

[2] 现在分词短语作定语

[3] 过去分词作定语,放在被修饰词的后面

[4] 定语从句

4.原文的参考译文

第5课　苹果在冷藏过程中农药残留的变化

采后生理病害是苹果长期贮藏的限制性因素。正如在引言中提到的，果园管理，如卫生、杀真菌剂使用以及害虫控制措施对于采后冷藏过程中潜在发生的腐烂类型和数量都有重要的影响。关于农作物的贮藏期限，当接近收获期时使用杀菌剂，也许可以控制损伤性病原体，其来源于果园中新近的真菌感染以及收获过程和采后处理中造成的伤口(刺破、擦伤等)真菌感染。然而，应该强调的是除了化学药物对作物保护所得到的有利之处，与使用杀虫剂所关联的身体危害也必须考虑进去。为了满足消费者和毒理学家的关注，必须控制食品供应中有可能发生的农药残留。我们前期研究了苹果预采收期中农药残留的动力学，也研究了直接供应给消费者和(或)婴幼儿食品加工的水果的最低农药残留的处理措施。本研究中，我们主要着眼于苹果在采后冷藏条件下残留农药的分解。在采收期对经FT1、FT2、FT3田间试验的果园苹果进行农药残留检测的结果如图1所示(田间试验FT4中，用于苹果树处理的杀虫剂在采收期都没有留下可检测的农药残留)。用于苹果树保护的杀虫剂的21个活性成分，于贮藏的开始阶段的苹果中发现了只有6种杀真菌剂和1种杀虫剂。啶虫脒、甲基氯吡磷、苯醚甲环唑、除虫脲、二氰蒽醌、二硫代氨基甲酸盐(在本研究中代表代森锰锌和福美双)、苯氧威、醚菌酯、伏虫隆、噻虫啉、唑蚜威、戊菌酯、杀虫脲的残留都降低到本研究中所采用分析方法的检测限(见表5，各个被分析化合物的检测限)。进一步的农药残留降低发生在冷藏阶段，结果见图2。在其中的一项试验中，5个月之后，在那些被检测的真菌杀菌剂中仅发现了多果定，而有机磷农药具有高度的持久性(见图2)。在这两种情况下，农药残留均低于0.01 mg/kg，这是原料允许用于婴幼儿产品加工的最大残留值。在下面的文章中，介绍了有关我们在采后期所检测到的杀虫剂更详细信息。值得注意的是，结果选择了代表不同化学类别的化合物，这些化学物质通常用于一些传统的苹果园中，也是监测体系中对成熟苹果做的常规的检测项目。

作　业　汉译英

1.试验主要研究了焙焦对荞麦麦芽的α-淀粉酶、β-淀粉酶(总的和可溶性的)、β-葡聚糖酶以及蛋白酶活性的影响。将普通荞麦在10 ℃下浸麦12 h，在15 ℃下发芽4 d，在40 ℃下焙焦48 h。在焙焦过程中分析了水分含量和酶活性。结果发现经过48 h的焙焦，水分含量从44%降为5%。

2.本文研究了在成熟过程中白色番石榴和粉色番石榴的果肉和果皮中组成成分的变化。白色番石榴和粉色番石榴是典型的呼吸跃变型果实。两种类型的番石榴相似，都是随着果实的成熟，果实硬度不断降低。随着果实硬度的降低，两种类型的番石榴的果肉和果皮中总可溶性固形物含量和总糖含量增加。

3.采用4种不同的方法(ABTS，DPPH，DMPD和FRAP)评价石榴汁的抗氧化活性，并与红酒和绿茶汤进行了比较。商品化的石榴汁的抗氧化活性(18～20 TEAC)是红酒和绿茶

(6～8 TEAC)的 3 倍。由整个石榴得到的商品化石榴汁的抗氧化活性比仅仅从果肉上得到的石榴果汁要高。

4.一般来说,热加工几乎可杀死所有的微生物,并可钝化果胶酶的活性,抑制微生物的产生,还使番茄汁有较好的稳定性。然而,同新鲜的汁液相比,热加工降低了产品的色泽、类胡萝卜素和番茄红素的萃取率以及维生素 C 的含量。在贮藏过程中,同新鲜番茄汁相比,所有的压力加工工艺能够提高类胡萝卜素和番茄红素的萃取率,同时比热加工保留了更多的维生素 C。

5.展青霉素是苹果和苹果产品中的一种重要的毒菌素,也是苹果和苹果汁工业中的质量标志。大量文献报道了有关在澄清和混浊的苹果汁以及液化的苹果块中萃取和分析展青霉素的研究。然而,在干的固体苹果产品中缺乏有关展青霉素分析的信息,如苹果干就不能被液化。我们研究了一种方法解决了这一问题,并验证了该方法的精密度、精确度和线性分别为 10×10^{-9}、30×10^{-9} 和 50×10^{-9}。这种方法是建立在液相色谱与二极管列阵检测器联用仪的固相萃取和等强度分离基础之上的。

第 8 章

论文的结论

Conclusion of Article

第 1 课　Rheological behaviour of dairy products as affected by soluble whey protein isolate

1.原文

选自： International Dairy Journal，2006，16：399-405

作者： George Patocka，Radka Cervenkova，Suresh Narine，Paul Jelen

CONCLUSION

The addition of soluble whey protein powder to liquid yoghurt and buttermilk lowered the product viscosity. The amount of whey protein affected the magnitude of the viscosity lowering effect，with the lowest viscosities recorded for 6%-10% WP addition，despite the significant increase in total solids content. Further increase of whey protein level was followed also by gradual return of the product viscosities to the control levels. The concentration dependence of the apparent viscosity on shear rate，[1] described by other researchers for heat treated whey protein systems，was confirmed by this study also for undenatured whey proteins. Three areas of fluid characteristics were identified；low concentration zone（<10%）with Newtonian characteristics；concentration between 10% and 35% with gradual development of pseudoplastic characteristics；and shear-thinning characteristic when concentration of whey protein exceeds 35%. The amount of whey protein and the moment of its addition affected the rheological and textural attributes of the stirred yoghurts. The addition of up to 10% WPI to stirred yoghurts[2] containing stabilizers produced a thinning effects with progressively decreasing G' . Incorporation of WPI <6% in experimentally prepared unstabilized yoghurt[3] added before fermentation produced changes[4] resulting in physical properties similar to the commercial product. WPI addition after fermentation caused breakdown of the unstabilized yoghurt into two phases，likely due to the mechanical effect of mixing combined with the redistribution of water upon addition of the soluble whey proteins. A similar effect was observed in the BF yoghurts with WPI content above 6%，[5] indicating permanent disruption of the casein-β-lactoglobulin gel texture[6] formed during fermentation of the heated yoghurt base，in contrast to the stabilizing effects of hydrocolloids[7] used in the commercial yoghurt.

2.词或词组

BF（before fermentation）　发酵前

whey　*n*.　乳清

buttermilk　*n*.　酪乳，白脱牛奶

denature　*v*.　变性

G'（storage moduli）　贮能模量 G'

hydrocolloid *n*. 水状胶质,水状胶体

lactoglobulin *n*. 乳球蛋白

magnitude *n*. 量,量级,大小

rheological *adj*. 流变学的

pseudoplastic *adj*. 假塑性的

shear *v*. 剪切

stabilizer *n*. 稳定器,稳定剂

viscosity *n*. 黏度,黏性

WPI (whey protein isolate) 分离型乳清蛋白

yoghurt *n*. 酸奶,酸乳酪

3.主要语法现象

[1][3][6][7] 过去分词短语作后置定语

[2][4]现在分词短语作定语

[5] 现在分词短语作独立成分

4.原文的参考译文

第 1 课 可溶性的乳清蛋白对乳制品的流变学性质的影响

结论

将可溶性的乳清蛋白粉添加到液态酸奶和酪乳中可以降低产品的黏度。乳清蛋白的用量影响了黏度降低的程度,在乳清蛋白添加量为 6%~10%时黏度最低,尽管此时固形物的含量显著增加。乳清蛋白的进一步增加将会使产品的黏度逐渐回升到可控制水平。其他研究人员所描述的在热处理的乳清蛋白体系中外观黏度所对应的浓度依赖于剪切率的结论,在本次使用未变性的乳清蛋白研究中也得到了证实。三个具有不同流体特性的区域被确定了:牛顿学说特点的低浓度区(10%)、浓度介于 10%~35%之间的逐渐发展的假塑性特点区域以及当浓度超过了 35%之后的剪切变稀特点区域。乳清蛋白的添加量和添加时刻影响了搅拌型酸奶的流变学特性和组织特性。添加 10%的分离型乳清蛋白到含有稳定剂的搅拌型酸奶中,产生了一个稀释的影响并伴随着贮能模量 G' 的逐渐下降。而在发酵前,添加小于 6%的分离乳清蛋白到未经稳定处理的酸奶中将产生一些变化,这些变化导致产品的物理特性相似于商业产品的物理特性。在发酵后添加乳清蛋白将导致未经稳定处理的酸奶分成两相,可能是由于混合以及添加了可溶性乳清蛋白后水分的重排而产生的机械效应所致。在发酵前添加了大于 6% WPI 的酸奶中,也观察到了相似的现象,说明在热酸奶发酵期间形成的酪蛋白-β-乳球蛋白凝胶结构的永久破坏,与用于商业酸奶中的水状胶体的稳定效应是截然相反的。

第 2 课　Effect of ethylene in the storage environment on quality of ‘Bartlett pears’

1. 原文

选自： Postharvest Biology and Technology，2003，28：371-379

作者： Bower J H，Biasi W V，Mitcham E J

CONCLUSION

These results demonstrate that temperature control is a more important factor in[1] maintaining pear quality than[2] scrubbing ethylene. While it is certainly desirable[3] to minimize ethylene concentrations around stored pears in order to reduce the incidence of scald and internal breakdown，from a practical standpoint it is important[4] that this is not done at the expense of good temperature control. There was little difference between pears[5] exposed to 1 μL/L and those exposed to 10 μL/L，[6] suggesting[7] that very low concentrations would have to be achieved for there to be any benefit in terms of improved quality. The high rate of ethylene production by the fruit made it difficult[8] to maintain ethylene concentrations at = 1 μL/L，especially at 2 ℃. In a commercial situation，where large amounts of fruit are packed tightly into cool stores，it is likely[9] to be extremely challenging if not impossible to keep ethylene to such low levels. Softening，yellowing，scald development and internal breakdown differed greatly between −1 and 2 ℃. Storage life of ‘Bartlett pears’ stored at 2 ℃ is clearly less than the 3 months[10] used in this trial，whereas fruit held at −1 ℃ ripened normally after this time. Further work is necessary[11] to determine the critical temperature for[12] storing pears above which excessive ripening and physiological damage will occur during long term storage. The interaction between temperature and the effects of ethylene should also be further defined，as it is clear that the effects of ethylene are quite different between temperatures as close as −1 and 2 ℃.

2. 词或词组

Bartlett pear　巴特利特梨（巴梨），香蕉梨（西洋梨的一种）

climacteric fruit　呼吸跃变型水果

ethylene　*n.*　乙烯

scrub　*v.*　洗擦，擦净，使（气体）净化，擦洗；　*n.*　洗擦，擦净，洗擦，擦净

scald　*n.*　烫伤，烫洗；　*v.*　被热油烫伤手，以沸水或蒸汽清洗，烫伤

standpoint　*n.*　立场，观点

softening　*n.*　软化，变软

3. 主要语法现象

[1][2][12] 现在分词作定语

[6] 分词短语作非限制性定语

[3][8][9][11] 不定式结构作主语
[4] 强调句
[5][10] 过去分词作定语
[7] 宾语从句

4.原文的参考译文

第2课 贮藏环境中乙烯对巴特利特梨质量的影响

结论

这些结果表明,对比乙烯消除法,温度控制是保持梨品质的一个更为重要的影响因素。但为了降低梨的褐斑和内部溃烂,我们的确想要将贮藏梨周围的乙烯浓度降至最低。从实际的角度来看,有了合理的温度控制,就不必降低乙烯浓度了。将梨暴露在1 μL/L和10 μL/L乙烯下没有什么区别,说明想有效改善梨的贮藏质量,就必须要达到非常低的浓度。梨中乙烯的高产率使得维持乙烯浓度低于1 μL/L非常困难,特别是在温度为2 ℃时。在商业环境中,大量的水果被紧密地堆放在冷库中,要保持如此低的乙烯浓度可能是非常具有挑战性的。软化、变黄、褐斑的发生和内部溃烂在－1～2 ℃之间是有很大不同的。在2 ℃下,巴特利特梨贮藏货架期明显短于3个月(本实验中的货架期),而贮藏在－1 ℃下,梨的正常成熟就延后了。进一步决定贮藏梨的临界温度的研究还需要继续进行,高于这个温度长期贮藏必将导致梨的过度成熟和生理损害。温度和乙烯二者之间相互作用的影响也应该进一步详细说明,因为很显然温度在接近－1～2 ℃时乙烯的影响完全不同。

第3课 Effect of nisin on yogurt starter, and on growth and survival of *Listeria monocytogenes* during fermentation and storage of yogurt

1.原文

选自: Internet Journal of Food Safety, 2003, 1: 1-5

作者: Benkerroum N, Oubel H, Sandine W E

CONCLUSION

These results confirm[1] that food processors should not solely rely on pasteurization and fermentation[2] to insure full protection and safety of fermented milk products. Some food grade additives may also be used in addition along with Good Manufacture Practices. Nisin proved to be effective in[3] controlling the growth of *L. monocytogenes* in both cottage cheese and yogurt (the present study). For the latter, the amount of nisin[4] used should inhibit the pathogen without[5] harming significantly the yogurt starter. In our case, a nisin concentration of up to 50 RU/mL had no adverse effect on the yogurt starter[6] we used (e. g Rediset). Moreover, nisin addition to yogurt may play a beneficial role in[7] keeping yogurt acidity from dropping too low during storage and thus prevents deterioration of its sensory attributes and extends its shelf-life. Such use is of more interest in developing countries[8] where yogurt is rarely stored at refrigeration temperature when marketed. Nisin (50 RU/mL) was added to reconstituted NDM previously adjusted to different pH values along with *L. monocytogenes* ATCC 7644 (ca., 10 ccll/mL). The experiment was conducted against a positive (only *Listeria* was added) and a negative (without added nisin or *Listeria*) controls. Growth of *L. monocytogenes* ATCC 7644 was monitored at 37 ℃ by plate count on TSA at 0, 4, 7, 24 and 48 hours.

2. 词或词组

cottage cheese 农舍奶酪,农家干酪
deterioration *n.* 变坏
Good Manufacture Practices 良好操作规范
inhibit *v.* 抑制
Listeria monocytogene (LM) 单核细胞增生李斯特菌
nisin *n.* 乳酸链球菌素
NDM (nonfat dry milk) 脱脂奶粉
pasteurization *n.* 加热杀菌法,巴氏杀菌法
reconstituted *adj.* 再生的,再造的
TSA(trypticase soy agar) 胰蛋白酶大豆琼脂

3. 主要语法现象

[1] 宾语从句
[2] 不定式结构
[3][5][7] 分词短语作介词宾语
[4] 过去分词作定语
[6] 省略 that 的定语从句
[8] 定语从句

4. 原文的参考译文

第3课 在酸乳发酵和贮藏期中乳酸链球菌素对酸奶发酵剂和单核细胞增生李斯特菌生长的影响

结论

这些结果表明,食品生产者不应该只依赖于巴氏消毒和发酵来完全地保证发酵乳制品的安全,除了良好操作规范外,也许应该使用一些食品级的添加剂。本实验证明,乳酸链球菌素在控制农舍奶酪和酸奶制品中的单核细胞增生李斯特菌方面是有效的。对于后者,乳酸链球菌素的用量应该控制在可抑制病原菌生长,但对酸奶发酵剂不产生显著性影响的水平。在我

们的研究中，乳酸链球菌素最高浓度50 RU/mL，对我们使用的酸奶发酵剂没有不利的影响。另外，在酸奶中添加乳酸链球菌素，也许有利于避免酸奶在贮藏期间酸度过度降低，这可防止酸奶感官品质的退化，并延长其货架期。这种添加剂的使用更适合发展中国家，因为在这些国家的卖场中很少将酸奶置于冷藏条件下保存。本实验中，将乳酸链球菌素（50 RU/mL）加入之前调了不同pH的包含有单核细胞增生李斯特菌ATCC 7644的再造脱脂奶粉中，还做了一个阳性对照（只加李斯特菌）和一个阴性对照（没加乳酸链球菌素或者李斯特菌）。控制单核细胞增生李斯特菌ATCC 7644在37 ℃下生长，并用胰蛋白酶大豆琼脂平板分别培养0、4、7、24和48 h，进行菌落计数。

第4课 Effect of thermal blanching and of high pressure treatments on sweet green and red bell pepper fruits (*Capsicum annuum* L.)

1.原文

选自：Food Chemistry，2008，107：1436-1449

作者：Castro S M，Saraiva J A，Lopes-da-Silva J A，Delgadillo I，Loey A V Smout C，Hendrickx M

CONCLUSION

Pressure treatments[1] applied to green and red peppers (100 and 200 MPa for 10 and 20 min) caused a lower reduction on soluble protein and ascorbic acid contents, than thermal blanching, particularly for red peppers,[2] that showed even an increase in the amount of ascorbic acid content, compared to the unprocessed peppers. Both green and red pressure treated peppers showed a level of residual polyphenol oxidase activity similar of that of thermally blanched peppers. Pectin methylesterase activity was only detected in green peppers and its activity declined progressively,[3] as blanching temperature and time increased, while the pressure treatments caused a slight increase of its activity. Peroxidase was more stable to the pressure treatments, than to the blanching treatments, particularly for red peppers, and showed a lower stability to the thermal blanching treatments than polyphenol oxidase and pectin methylesterase. Thermal blanching and pressure treatments caused at maximum 1-2 decimal reductions on microbial load,[4] pointing out the importance of the washing step with chlorine solution for the microbial quality of the final product.

Firmness was equally to better retained in pressure treated peppers[5] compared to thermally blanched peppers, before and after freezing, while red peppers showed higher sensitivity to lose firmness, compared to green peppers. Firmness was more affected by freezing when it was measured from the flesh side. Globally, the pressure treated peppers present similar to better levels of the quality parameters studied,[6] pointing to the possible use of pressure treatments as an alternative to the conventional thermal blanching of sweet bell peppers. Pressure treatments at higher pressures should be tested in further studies,[7] to evaluate if they can cause higher reductions in microbial loads and enzymatic activity, and still yield peppers with better nutritional and texture characteristics, compared to blanched peppers.

2.词或词组

bell pepper　柿子椒
thermal　*adj*.　热的,热量的
blanching　*n*.　热烫
polyphenol　*n*.　多酚
oxidase　*n*.　氧化酶
pectin　*n*.　胶质,果胶
methylesterase　*n*.　甲基酯酶
peroxidase　*n*.　过氧化物酶,过氧化氢酶
decimal　*adj*.　十进制的
chlorine　*n*.　氯

3.主要语法现象

[1][5] 过去分词作定语
[2] 定语从句
[3] 状语从句
[4][6] 分词短语作定语
[7]不定式结构

4.原文的参考译文

第 4 课　热烫和高压处理对青椒和红椒(*Capsicum annuum* L.)的影响

结论

与热烫法相比,用压力处理法处理青椒和红椒(100 和 200 MPa 分别处理 10 和 20 min),能够使可溶性蛋白和抗坏血酸含量有一个较低的下降,特别是红椒,比起未处理过的辣椒,甚至可使抗坏血酸的含量增加。经压力处理的青椒和红椒中残留的多酚氧化酶,与经热烫处理的辣椒相比,显示出了相似的活力水平。果胶甲基酯酶的酶活力只在青椒中测得,并且它的活力随着热烫温度和时间的增加不断下降,而压力处理法却使得它的活力略微增加。过氧化氢酶在压力处理法中比热烫法中更稳定,特别是红椒,在热烫法中,该酶的稳定性低于多酚氧化酶和果胶甲基酯酶。热烫和压力处理最大可将微生物数量减少 1～2 倍,说明用含氯的溶液清洗对于最终产品微生物质量的重要性。

在冷冻前后,压力处理的柿子椒和热烫处理的柿子椒硬度保持的一样好,而红椒比青椒对

于硬度的丧失有更高的敏感性。当从果肉一边测定硬度的时候,硬度更易受到冷冻的影响。从全球来看,经压力处理的辣椒所呈现出的质量水平,与以前研究得出的较好的质量参数水平相似。这些研究指出了压力处理法可替代传统的热烫法应用于甜柿子椒,将来应进行更高压力处理的研究,以评价更高的压力是否能导致微生物数量和酶活力的进一步下降,并且与热烫相比,压力法处理过柿子椒依然能够保持更好的营养和质地特点。

第5课 Gelatine-starch films:physicochemical properties and their application in extending the post-harvest shelf life of avocado (*Persea americana*)

1.原文

选自: Journal of the Science of Food and Agriculture, 2008,88: 185-193

作者: Miguel A. Aguilar-Méndez, Eduardo San Martín-Martínez, Sergio A. Tomás, Alfredo Cruz-Orea, Mónica R. Jaime-Fonseca

CONCLUSIONS

The CO_2P was significantly dependent on the starch concentration in the film;[1] when increasing the content of this polysaccharide the CO_2P values also increased. The lowest CO_2P value [4.35×10^{-15} kg/(ms·Pa)] was obtained at 2 g/kg of starch concentration, 6.5 g/kg of glycerol and pH 6. With respect to film PS, higher values were obtained with a glycerol content of 4 g/kg, for the different pH values (3.5, 6 and 8.5). On the other hand, film deformation values were higher when low starch concentration and high glycerol content were used in the preparation of the plasticized films. The applicability of these films was evaluated in avocado fruits. Under low temperature conditions, coated fruits showed lower weight loss, better surface colour and higher pulp firmness as compared with the control avocados. In addition, the respiratory climacteric pattern of coated fruits,[2] stored at 20 ℃, was delayed for 3 days. The coatings also improve the fruit appearance and had a good stability at low temperature (6 ℃). From these results, it was possible to conclude[3] that the use of gelatine-starch coatings increased the shelf life of 'Hass' avocados[4] preserving their fresh-like characteristics.

2.词或词组

avocado *n.* 鳄梨

climacteric *n.* 转变期,呼吸期,更年期

coating *n.* 涂膜,涂层

CO_2P (carbon dioxide permeability) 二氧化碳渗透率

film *n.* 膜

gelatine *n.* 胶质,白明胶

Hass avocado 汉斯鳄梨(世界著名鳄梨)

glycerol　*n*.　甘油，丙三醇

polysaccharide　*n*.　多糖

PS (puncture strength)　戳穿强度

plasticize　*v*.　使塑化

pulp　*n*.　果浆，果肉

respiratory　*adj*.　呼吸的

3.主要语法现象

[1] 状语从句

[2] 过去分词作定语

[3] 宾语从句

[4] 现在分词短语作同位语

4.原文的参考译文

第 5 课　明胶-淀粉膜的理化性质及其在延长鳄梨采后货架期方面的应用

结论

二氧化碳渗透率(CO_2P)显著地依赖于膜的淀粉浓度，当增加多糖含量时，二氧化碳渗透率也随之增加。在淀粉浓度为 2 g/kg、甘油浓度为 6.5 g/kg 且 pH 为 6 时，CO_2P 值最低[4.35×10^{-15} kg/(ms·Pa)]。当甘油浓度为 4 g/kg，不同的 pH(3.5、6 和 8.5)时，膜的戳穿强度(PS)较高。另一方面，当制备可塑性膜时，低浓度的淀粉和高浓度甘油会使膜的变形值较高。膜的可用性可通过鳄梨来进行评价，在低温条件下，比起对照组的鳄梨，涂膜的水果显示了较小的质量损失，较好的表面颜色和较高的果肉硬度。除此之外，涂膜水果的呼吸跃变模式，在 20 ℃的贮藏条件下被延后了 3 d。涂膜也改善了水果的外观，并在低温下(6 ℃)有较好的稳定性。从这些结果中可能得出结论，那就是胶质淀粉涂膜可以增加汉斯鳄梨的货架期，并保持其新鲜特性。

第 6 课　Influence of cooking and microwave heating on microstructure and mechanical properties of transgenic potatoes

1.原文

选自： Nahrung/Food，2004，48：169-176

作者： Blaszczak W，Sadowska J，Fornal J，Vacek J，Flis B，Zagórski-Ostoja W

CONCLUSION

(ⅰ)Genetic modification of potatoes towards virus PVY showed to have an influence on variability of mechanical properties of potato tubers within and between clone groups, but no simple correlation was found. (ⅱ)The[1] observed differentiation in the mechanical properties of heat-treated potatoes was less connected with modification on the genetic level but most of all with a kind of the process used. (ⅲ)The effect of thermal processing (cooking and microwave heating) of potato tubers on their mechanical properties was evident, independently of genetic modification. (ⅳ)Microwave heating of potato tubers caused a higher decrease in mechanical resistance of tubers than conventional cooking process. (ⅴ)Deformation of parenchyma cells during cooking was directly connected with starch gelatinization and gel formation. (ⅵ) Microwave heating affected significantly cellular water evaporation[2] which resulted in intercellular failure, collapsing of cells, and limitation of starch gelatinization.

2.词或词组

deformation *n.* 变形
gelatinization *n.* 胶凝(作用)
intercellular *adj.* 细胞间的
ready-to-eat 方便食品,即食食品,速煮食品
parenchyma *n.* 软组织
PVY (potato virus Y) 马铃薯病毒 Y
thermal *adj.* 热的,热量的

3.主要语法现象

[1] 过去分词作定语
[2] 定语从句

4.原文的参考译文

第 6 课 烹调加热和微波加热对转基因马铃薯微结构和机械特性的影响

结论

(1)针对马铃薯病毒 Y 所做的马铃薯基因修饰显示了它对克隆组内和组间的马铃薯块茎的机械特性的变化有影响,但是没有发现它们之间存在相关性。(2)在热处理的马铃薯中观察到的机械特性的区别与基因水平的修饰关系不大,而与所采用的加工过程有很大关联。(3)热加工(烹调和微波加热)对马铃薯机械特性的影响是明显的,与基因修饰无关。(4)与传统的烹调加工相比,微波加热使块茎的抗机械性有了很大的下降。(5)在烹制过程中软组织细胞的变形与淀粉的糊化作用及凝胶的形成有直接关系。(6)微波加热严重影响了细胞水分的蒸发,这导致了细胞间分裂和细胞破坏,并限制了淀粉的糊化作用。

第 7 课　Effects of drying process on antioxidant activity of purple carrots

1. 原文

选自： Nahrung/Food 2004, 48: 57-60

作者： Seda Ersus Uyan, Taner Baysal, Ünal Yurdagel, Sedef Nehir El

CONCLUSION

The use of DPPH provides an easy and rapid way to evaluate the antiradical activities of antioxidants. The EC_{50} value of purple carrot has not been mentioned before in the literature. We aimed to measure the purple carrot antioxidant capacity by DPPH method and the effects of different drying methods on the anthocyanin content and EC_{50} value. The consumption of a more concentrated source of antioxidant of purple carrot was produced. The total drying times were found similar for tray drier (150 min) and for microwave (45 min) + tray drier (105 min). So for further studies, it will be better to try only the microwave drying process to the purple carrots to decrease the drying time effectively. Also the blanching step can be eliminated to inhibit the loss of water-soluble compounds in blanching water. No browning problem of purple carrots during processes occurred.

2. 词或词组

DPPH(1,1-diphenyl-2-picrylhydrazyl)　1,1-二苯基-2-苦肼基

EC_{50}　降低 50% 初始 DPPH 浓度所需的样品量

antiradical　*adj*.　抗自由基的

antioxidant　*n*.　抗氧化剂

anthocyanin　*n*.　花青素

catechin　*n*.　儿茶酚

eliminate　*v*.　排除，消除

epicatechin　*n*.　表儿茶酸

flavone　*n*.　黄酮

isoflavone　*n*.　异黄酮

purple carrot　紫胡萝卜

tray drier　盘式干燥器

3.原文的参考译文

第7课　干燥过程对紫胡萝卜抗氧化活性的影响

结论

1,1-二苯基-2-苦肼基(DPPH)的使用提供了一个简便而快速的评价抗氧化剂的抗自由基活性的方法。以前的文献没有提到过紫胡萝卜的 EC_{50} 值。我们的目的是通过 DPPH 法测定紫胡萝卜的抗氧化能力,以及不同的干燥方法对花青素含量和 EC_{50} 值的影响,即可计算提取紫胡萝卜的抗氧化剂浓缩物的损失量。用盘式干燥器(150 min)的干燥时间和用微波(45 min)+盘式干燥器(105 min)的总干燥时间是相似的,进一步研究发现,最好用微波干燥紫胡萝卜,可以有效地减少干燥时间。应省去热烫的工序,以减少水溶性成分在热水中的损失。在加工过程中未发生紫胡萝卜褐变的问题。

第8课　Genetically improved starter strains: opportunities for the dairy industry

1.原文

选自: International Dairy Journal, 1999,9: 11-15

作者: Beat Mollet

CONCLUSION

The above-mentioned examples illustrate[1] the vast potential lactic acid bacteria have for present and future applications in various domains of the dairy industry. They will continue to play an important role in the fermentation processes of a variety of different food products by[2] contributing to their conservation, flavour development, texture and health beneficial properties. Increasingly they will also be used as a natural source of food ingredients and additives to non-fermented products to attain for example organoleptic, texturing or probiotic aims. The progress in molecular biology and genetic engineering will broaden the possibilities for[3] using lactic acid bacteria in food and may also allow the improvement of existing products and the development of novel products and applications.

Lactic acid bacteria have a history of safe use in fermented food products. They are considered non-toxic and have not been reported as the causative agents in disease or infection. Nevertheless, all newly isolated or genetically altered lactic acid bacteria and their resultant fermented foods have to be carefully evaluated for their safety to the consumer and environment. They are subjected to stringent safety and regulatory procedures in the same way[4] as do all new food products[5] which are commercialized and sold to the consumer. In fact, it was only with the help of molecular biology[6] that it became possible to correctly classify and differentiate the diverse species of the lactic acid bacteria to better identify their origins, natural habitats and dissemination in nature and in the intestines of humans and animals. This knowledge and the possibility to trace individual strains along the food chain and the digestive tract further contributes to the better evaluation of potential risk factors and finally adds to their confidence and safety. The intelligent, responsible development and use of genetically improved microorganisms will lead us into the new millennium[7] bringing exciting new products to the consumer.

2.词或词组

causative *adj*. 有问题的,成为原因的
digestive tract 消化道
diverse *adj*. 不同的,变化多的
dissemination *n*. 分布,分发
infection *n*. 传染,感染
intestine *n*. 肠道
millennium *n*. 千年期,千周年纪念日
probiotic *n*. 益生菌

3.主要语法现象

[1] 省略 that 的宾语从句
[2][3][7] 分词短语
[4] 状语从句
[5][6] 定语从句

4.原文的参考译文

第8课 基因改造的酵母菌株:乳制品工业的机遇

结论

以上提到的例子阐明了这种具有巨大潜能的乳酸菌无论现在还是将来都可以用于乳制品工业的各个领域。因为用它们生产的乳制品在保藏、香味改善、质地和健康方面均表现出很好的特性,所以它们将继续在各种产品的发酵过程中扮演重要角色。它们也将作为一种食品成分和天然来源的添加剂越来越多地用于非发酵产品,以达到改善感官品质,增强结构特性以及作为益生菌的目的。分子生物学和基因工程的发展与进步将增加乳酸菌在食品中应用的可能,并且可能改进现有产品,发展出新产品和新的应用。

乳酸菌在发酵食品中有着安全的使用史。它们被认为是无毒的,而且没有任何导致疾病或传染病的报道。然而,所有的新分离出的乳酸菌,或者是改变了基因的乳酸菌,以及它们衍生出的发酵产品都必须经过对消费者和环境的细致的安全评估。它们和所有新的商业化的食品一样,都要受制于严格的安全规范程序。事实上,它是唯一在分子生物学的帮助下能够进行正确地分类和区别不同种属的微生物,进而能更好地识别它们的来源、天然栖息地,以及在自

然界和人类与动物肠道中的分布。这种沿着食物链与消化道去追踪单个菌种的可能性，将有益于进一步对潜在的危险因素做出更好的评价，最终增加人们对它们的信心和安全感。知识的逐步发展和基因改良微生物的应用必将带来令消费者兴奋的新产品，引领着我们走进新的千年。

第9课 Genetically engineered foods and the environment: a catastrophe in the making

1.原文

选自： Food Safety Review，2002，3：1-6

作者： Centre for Food Safety

CONCLUSION

Despite the misleading claims of companies selling them, GE crops will not alleviate traditional environmental concerns, such as the chemical contamination of water, air, or soil. Far from eliminating pesticides, GE crops may well increase this chemical pollution. Plants[1] engineered to tolerate herbicides closely tie crop production to increased chemical usage. Crops[2] engineered with Bt genetic material to protect against specific insect pests may decrease the efficacy of this important non-chemical pesticide by increasing resistance to it. This could mean the end of organic agriculture as we know it, with the conversion of this sustainable method of farming to chemical-intensive methods. Meanwhile, genetic engineering has brought an entirely new slate of environmental concerns. Altered genes[3] engineered into commercial plants are escaping into populations of weeds and unaltered crops. Genetically enhanced super weeds may well become a severe environmental problem in coming years. Even now, GE corn, canola and, to a lesser extent, soybeans and cotton are contaminating their non-GE counterparts. This is causing major economic concerns among farmers and is resulting in the loss of U. S. agricultural exports. The biological pollution[4] brought by GE crops and other organisms will not dilute or degrade over time. It will reproduce and disseminate, profoundly[5] altering ecosystems and threatening the existence of natural plant varieties and wildlife. Despite these troubling and unprecedented environmental concerns, the U. S. government has allowed companies to grow and sell numerous gene altered crops. In the U. S., more than 76 million acres are now planted with GE varieties. Yet no government agency has thoroughly tested the impact of these crops on biodiversity or farmland and natural ecosystems. No federal agency has ever completed an Environmental Impact Statement on any GE organism, and much research into the environmental impacts of GE crops remains to be done. No regulatory structure even exists to ensure that these crops are not causing irreparable environmental harm. The FDA,[6] our leading agency on food safety, requires no mandatory environmental or human safety testing of these crops whatsoever. Nonetheless, officials at the FDA, EPA, and USDA have allowed, and even promoted, GE crop plantings for years. The lack of government oversight is troubling. Each decision to introduce these biological contaminants into our environment is a dangerous game of ecological roulette. The extent of irreversible environmental damage grows greater with every new acre of GE cropland and every new GE variety.

2. 词或词组

biodiversity *n*. 生物多样性
disseminate *v*. 散布,传播
EPA(Environmental Protection Agency) 美国环保署
FDA(Food and Drug Administration) 美国食品和药品监督管理局
GE(genetic engineering)crop 基因工程农作物,基因改良作物,转基因作物
herbicide *n*. 除草剂
pesticide *n*. 杀虫剂
unprecedented *adj*. 空前的
USDA(United States Department of Agriculture) 美国农业部

3. 主要语法现象

[1][2][3][4] 过去分词作定语
[5] 现在分词短语
[6] 现在分词短语作同位语

4. 原文的参考译文

第 9 课 基因工程食品与环境:正在形成的灾难

结论

尽管销售转基因农作物的公司发布了一些误导人的宣传,但是并不能减少一直以来人们对环境问题的关注,如水、空气和土壤的化学污染。且不说消除虫害,转基因农作物也许会增加化学污染。经过基因改良耐除草剂的农作物的产量与通过增加化学药品用量而获得的产量是一样的。为了对抗特定的病虫害而应用 BT(具有抗虫性的土壤菌)的转基因作物,对病虫害的抗性提高,但是它抗非化学杀虫剂的能力可能降低。随着这种可持续性耕作方式向化学密集型方式的转化,可能意味着我们所熟知的有机农业时代的结束。同时,基因工程也把人们对于环境问题的关注带到了一个全新的层面。那些通过基因工程技术已在商业作物中发生改变的基因,也正偷偷地进入杂草和尚未改变基因的农作物当中,在未来几年,也许转基因超级杂草将成为一个严重的环境问题。即使是现在,转基因的玉米、油菜籽、大豆和棉花也在较小程度上污染着它们的非转基因作物,这也造成了农民主要的经济问题,并导致了美国政府农业出口的损失。这种由转基因作物和其他有机物带来的生物污染不能随着时间的推移而减少或消退。它们将繁殖和广泛传播,深深地改变生态系统,威胁自然植物种类和野生动植物的存

在。尽管存在这些令人不安和前所未有的环境问题，美国政府还是允许种植和买卖各种各样的转基因作物。在美国，超过 7 600 万英亩（1 英亩 = 4 047 m^2）的土地正在种植转基因作物，仍然没有任何机构彻底地测试过这些作物是否对于作物多样性或耕地和自然生态有影响。也没有任何联邦机构曾做过一份完整的关于基因工程菌的环境影响声明或报告。大量的关于转基因作物对环境影响的研究工作仍需继续进行。甚至没有仲裁组织来保证这些作物不会对环境产生不可修复的破坏。FDA（美国食品和药品监督管理局）是我们食品安全的主要机构，也不强制要求做任何这类作物对于环境和人类安全的实验。尽管如此，美国食品和药品监督管理局、美国环保署和美国农业部的官员们已经允许并推广种植转基因作物有数年之久了。政府监管的缺失是会容易引起混乱的。将生物污染引入环境过程中的每一个决定都是生态轮盘赌局中的一个危险游戏。环境的不可逆转的破坏程度将随着每英亩耕种转基因作物的农田以及每一个新的转基因作物品种的增加而日益加剧。

第 10 课　Evolution and stability of anthocyanin-derived pigments during port wine aging

1. 原文

选自： Journal Agriculture and Food Chemistry，2001，49：5217-5222

作者： Nuno Mateus，Victor de Freitas

CONCLUSION

The decrease in concentration of the three major anthocyanidin monoglucosides（mv，mv-ac，and mv-coum）and the anthocyanin pyruvic acid adducts（mv py，mv-ac py and mv-coum py）during three years of aging followed first order kinetics in all the wines studied. The pyruvic acid adducts were found to be very resistant to wine aging when compared to the original grape anthocyanins. The results of both anthocyanins and pyruvic acid adducts show[1] that acylation on the sugar moiety of all the pigments decreased their stability in wine. After wine spirit addition，it was shown that the levels of mv py and its acylated forms increased importantly before starting to decrease around 100 days.

The three major anthocyanins（which are precursors of the newly formed pigments）decreased almost totally during the same period. An initial increase of anthocyanin pyruvic acid adduct concentrations was concurrent with the degradation of anthocyanidin monoglucosides，although in a different range of concentrations. These studies[2] performed in wines are very complex due to their high chemical complexity，and the assessment of all factors[3] involved in the formation of new pigments is very difficult. Anthocyanin pyruvic acid adducts are more abundant in Port wines than in red table wines，as seen from previous analysis in our laboratories（data not shown）and as reported by other authors. The characteristic chemical and physical properties of Port wine could be at the origin of their higher levels of newly formed pigments. First，when wine spirit is added in order to stop fermentation, the pyruvic acid concentration is expected to be higher than when the fermentation is allowed to go to dryness. Effectively, the pyruvic acid[4] excreted by the yeast at the beginning of the fermentation is further used in the yeast metabolism. Beside this, Port wines present slightly higher pH values than red table wines,[5] which is an important factor that influences anthocyanin pyruvic acid adduct formation. Additionally, the higher content of ethanol,[6] which is known to be a good solvent for polyphenols, increases the pigment solubility and can probably favor the formation of new pigments. Finally, another important aspect is the chemical composition of the wine spirit used,[7] which represents about 20% of the final Port wine volume. It is known that wine spirit has a high level of aldehyde compounds, especially acetaldehyde, which favors the reaction between anthocyanins and tannins. Nevertheless, its real contribution is practically unknown. Further studies are thus needed regarding the influence of some of these factors in the evolution and stabilization of Port wine color.

2. 词或词组

acetaldehyde　*n.*　乙醛,醋醛
acylation　*n.*　酰基化
aldehyde　*n.*　醛,乙醛
anthocyanidin　*n.*　花青素
anthocyanin　*n.*　花青素苷
monoglucoside　*n.*　单葡糖苷
mv (malvidin 3-glucoside)　*n.*　二甲花翠素-3-葡萄糖苷,锦葵色素-3-葡萄糖苷
mv-ac (malvidin 3-acetylglucoside)　*n.*　二甲花翠素-3-乙酰葡萄糖苷
mv-coum (malvidin 3-coumaroylglucoside)　*n.*　二甲花翠素-3-香豆酰葡萄糖苷
Port　*n.*　波特(葡萄牙的港口城市)
polyphenol　*n.*　多酚
pyruvic acid　丙酮酸
tannin　*n.*　单宁

3. 主要语法现象

[1] 宾语从句
[2][3][4] 过去分词作定语
[5][6][7] 定语从句

4. 原文的参考译文

第 10 课　波特葡萄酒陈酿期间花青素类色素的稳定性和评价

结论

葡萄酒中的 3 种主要的花青素单葡萄糖苷(二甲花翠素-3-葡萄糖苷,二甲花翠素-3-乙酰

葡萄糖苷和二甲花翠素-3-香豆酰葡萄糖苷)的浓度和花青素丙酮酸聚合物(mv py,mv-ac py,mv-coum py)的浓度在陈酿过程中的减少,都遵循一级动力学规律。研究发现,相比于葡萄中原有的花青素,花青素丙酮酸聚合物更耐陈酿。两者的研究结果都显示,所有花青素中糖分的酰基化,都降低了它们在葡萄酒中的稳定性。在加入葡萄酒乙醇后,结果显示,二甲花翠素-3-葡萄糖苷丙酮酸聚合物(mv py)和它的酰基化形式的浓度在开始下降前约 100 d 时得到了显著的提高。

这 3 种主要的花青素(它们是新形成的色素的前体)几乎在同一时期完全减少了。花青素丙酮酸聚合物浓度的升高是和花青素葡萄糖苷的降解同时发生的,尽管浓度范围不同。由于葡萄酒具有很大的化学复杂性,所以这些研究是非常复杂的,因此对于新色素形成过程中影响因素的评估是非常困难的。和红葡萄酒相比,波特型葡萄酒含有更丰富的花青素丙酮酸聚合物,这从我们实验室先前的分析和其他作者的报道中都可以看到。波特型葡萄酒特有的化学和物理性质可能是其新色素含量高的原因。首先,加了乙醇终止发酵的葡萄酒液中的丙酮酸浓度,应该比不加乙醇一直发酵至终止的酒液中的丙酮酸浓度高,说明发酵初期由酵母分泌的丙酮酸,更进一步有效地用于酵母的新陈代谢当中。除此之外,波特型葡萄酒比红葡萄酒具有略高的 pH,这是影响花青素丙酮酸聚合物形成的一个重要因素。另外,高浓度的乙醇是多酚类物质的良好溶剂,可提高花青素的溶解性并且可能促成新花青素的合成。最后,另一个重要的方面是所加入的乙醇的化学成分,它约占波特型葡萄酒体积的 20%。众所周知,葡萄酒酒精含有大量的醛类成分,特别是乙醛,这有助于花青素和单宁之间的反应。尽管如此,实际上它真正的作用还不明确,这些因素对波特型葡萄酒颜色转变和稳定性方面的影响还需要进一步地研究。

第 11 课　Acrylamide in cereal products

1. 原文

选自: Journal of Cereal Science, 2008, 47(2): 118-133

作者: Achim Claus, Reinhold Carle, Andreas Schieber

CONCLUSION

Since the first detection of acrylamide in foodstuffs in 2002, significant progress in[1] understanding how acrylamide is generated has been made. Furthermore, many options and tools for reducing acrylamide in cereal products have been reported. In addition to changes in the temperature-time regime of baking and fermentation, replacement of crucial baking additives such as NH_4HCO_3 or invert sugar syrup also proved very effective. In addition, the use of additives such as cysteine, bivalent cations, or polyphenols seems to be very promising for acrylamide reduction in cereal products. However, strategies useful for potato products are not necessarily transferable to bakery products due to different limiting precursors and completely different product technologies. As a result, manufacturers will have to identify the most promising solutions for their respective products. This is evident[2] when gingerbread is compared with crispbread or bread rolls. Above all, manufacturers need to keep in mind consumer expectations regarding flavour, colour, and other sensory properties in order to ensure their products remain marketable.

The relative impacts of different formation mechanisms to total acrylamide levels also need to be assessed. This should help us to find suitable methods for its mitigation. In case of cereal products, more data from long-term agronomic studies are required to better understand and control the impact of the raw materials on acrylamide formation. It should be determined whether the type of fertiliser (calcium ammonium nitrate, ammonium sulphate, etc.) applied has an impact on amino acids and thus on acrylamide. Furthermore, the time of fertiliser application could play a crucial role. The most promising field for acrylamide reduction is the addition of low molecular additives such as polyphenols,[3] which have not so far been applied in cereal products. Nevertheless, trials with potato products indicate a high acrylamide-decreasing potential. Such additives ideally combine acrylamide reduction with little or no changes in product technology or, most importantly, sensory quality. Furthermore, possible health benefits from e. g. polyphenols could even enhance the consumer acceptance of such products.

2. 词或词组

acrylamide　*n.*　丙烯酰胺
bivalent　*adj.*　二价的
crispbread　*n.*　薄脆饼干
gingerbread　*n.*　姜饼
invert sugar syrup　转化糖浆
mitigation　*n.*　缓解，减轻
nitrate　*n.*　硝酸盐
sulphate　*n.*　硫酸盐

3. 主要语法现象

[1] 分词短语作状语
[2] 状语从句
[3] 定语从句

4. 原文的参考译文

第 11 课　谷类产品中的丙烯酰胺

结论

自从 2002 年在食品中首次发现丙烯酰胺后，人们在了解丙烯酰胺是如何产生方面有了很大的进步。而且，也出现了很多关于降低谷类产品中丙烯酰胺的方法和手段的报道。除了改变焙烤和发酵的温度-时间体系外，用碳酸氢铵或转化糖浆替代其他的焙烤添加剂也都被证明

是有效的。另外，添加半胱氨酸、二价阳离子或多酚类物质似乎对减少谷类中的丙烯酰胺也是非常有发展前景的。由于具有不同的限制性前体和完全不同的生产工艺，所以不能将适用于马铃薯产品的策略应用到焙烤食品中去。因此，生产商必须为不同的产品制定各自最有效的方法，当用姜饼和脆饼或面包圈进行比较时，这一点就非常明显。总之，生产商必须考虑消费者对于风味、颜色或其他感官品质的需求，这样才能让自己的产品在市场上占有一席之地。

各种不同的形成机制对总的丙烯酰胺水平的相对影响也有待评估，这有助于找到减少丙烯酰胺的方法。以谷类产品为例，我们需要得到更多长期农学研究的数据，以更好理解和控制原料对丙烯酰胺形成的影响。我们应该确定使用的肥料类型（硝酸铵钙、硫酸铵）是否对氨基酸有影响，进而影响了丙烯酰胺，而且肥料使用的时间也起着重要的作用。减少丙烯酰胺最有效的方法就是添加小分子的多酚类物质，但目前它还未应用在谷类产品中。然而，在马铃薯制品中的应用证明了它具有减少丙烯酰胺的潜在可能。这种添加剂可以减少丙烯酰胺，但对于产品工艺，特别是对感官质量有很小的影响，或者没有影响。而且，来自诸如多酚类物质对人体健康有益的作用也会增加消费者对此类产品的认可。

第12课 Nanoemulsion delivery systems for oil-soluble vitamins: influence of carrier oil type on lipid digestion and vitamin D_3 bioaccessibility

1.原文

选自：Food Chemistry，2015，187:499-506

作者：Bengu Ozturk，Sanem Argin，Mustafa Ozilgen，David Julian McClements

CONCLUSION

In this study, the influence of the type of carrier oil[1] used to formulate nanoemulsion-based delivery systems, on the bioaccessibility of an important oil-soluble bioactive (vitamin D_3) was examined. As observed with other highly lipophilic bioactive agents, it was found[2] that the nature of the carrier oil had a major influence on the bioaccessibility of vitamin D_3 measured using a simulated gastrointestinal model.

The rate and extent of lipid digestion was higher for MCT nanoemulsions than LCT nanoemulsions (corn oil and fish oil),[3] which was attributed to accumulation of long chain FFAs at the lipid droplet surfaces inhibiting lipase activity. As expected, indigestible lipids (orange oil and mineral oil) did not produce FFAs[4] when exposed to lipase. Vitamin bioaccessibility was higher in LCT nanoemulsions than in MCT nanoemulsions, presumably due to the higher solubilization capacity for vitamin D_3 of mixed micelles[5] formed by long chain FFAs. Surprisingly, a relatively high bioaccessibility for the nanoemulsions prepared from indigestible oils was found,[6] which may have been an experimental artefact associated with the presence of vitamin-containing small lipid droplets in the micelle phase. Alternatively, it may be possible for these small lipid droplets to be adsorbed by the human body, and therefore increase vitamin bioavailability. In summary, LCT nanoemulsions were found to be the most suitable for increasing the bioaccessibility of vitamin D_3. These results are important for formulating nanoemulsion-based delivery systems for oil-soluble vitamins and other lipophilic nutraceuticals.

2. 词或词组

bioaccessibility　*n.*　生物可接受度
bioavailability　*n.*　生物可利用度
droplet　*n.*　液滴
FFAs (free fatty acids)　游离脂肪酸
formulate　*v.*　规划,用公式表示,明确地表达
gastrointestinal　*adj.*　胃肠的
indigestible　*adj.*　难消化的
LCT (long chain triglycerides)　长链甘油三酯
lipophilic　*adj.*　亲脂性的,亲脂的
MCT (medium chain triglycerides)　中链甘油三酯
micelle　*n.*　胶束,微团
nanoemulsion　*n.*　纳米乳液
solubilization　*n.*　溶解,增溶

3. 主要语法现象

[1][5] 过去分词作定语
[2] 宾语从句
[3][6] 定语从句
[4] 状语从句

4. 原文的参考译文

第 12 课　脂溶性维生素的纳米乳液传递系统：载体油脂类型对油脂消化和维生素 D_3 生物有效性的影响

结论

本文研究了在纳米乳液传递系统中载体油脂类型对重要的脂溶性活性物质(维生素 D_3)生物有效性的影响。应用模拟胃肠道模型进行的研究发现,载体油脂性质对维生素 D_3 的生物有效性有重要的影响,这与用其他高度亲脂生物活性物质研究所观察到的结果类似。

与长链甘油三酯纳米乳液(玉米油和鱼油)相比,中链甘油三酯纳米乳液的消化速率更快,消化程度也更高,这是因为抑制脂肪酶活性的长链游离脂肪酸积聚在液滴表面。与预期结果一致,难消化的脂质(橙油和矿物油)在放入脂肪酶后没有被消化成游离脂肪酸。与中链甘油三酯纳米乳液相比,长链甘油三酯纳米乳液中的维生素的生物有效性更高,这可能是由于长链

游离脂肪酸形成的混合胶束对维生素 D_3 有更好的溶解作用。令人惊讶的是，以难消化的油脂为载体制备的纳米乳液却具有更高的生物有效性，这可能是一种实验假象，是由胶束相中存在有包裹维生素的小液滴而造成的。另一种可能是，这些小液滴可能被人体吸收了，从而提高了维生素的生物可利用度。总之，长链甘油三酯纳米乳液最适宜提高维生素 D_3 的生物有效性。上述结果对制备运载脂溶性维生素和其他亲脂营养制剂的纳米乳液传递系统十分重要。

作　业　汉译英

1.浓缩型乳清蛋白由于具有卓越的工艺性能和较高的营养价值，常被用于各种各样的食品当中。它们被当作结构成分用于液态食品中以增加硬度并促使热凝胶形成。除此之外，也有报道称乳清蛋白具有特殊的生理学特性。未变性的乳清蛋白现在正在被作为食品辅助剂来研究，通过细胞内的谷胱甘肽的合成来改善免疫状况。

2.乙烯在贮藏环境中的存在，会缩短许多果蔬的货架期，同时还会增加果蔬的腐烂、导致许多生理紊乱的发生。以呼吸跃变型水果为例，外源的乙烯可以加速成熟，但是这在贮藏水果中是不希望发生的。这暗示了贮藏环境中乙烯的浓度和许多果蔬的质量损失率有直接的关系。

3.因为食用了含有李斯特菌的食品而引发的疾病一直在唤起人们对于食品安全的关注。在美国，约25%因食品传播疾病而死亡的病例都是由单核细胞增生李斯特菌导致的。乳和乳制品则是最经常被控告的产品。因此，为了控制乳品中的病原菌，人们已经做出了相当大的努力。在乳制品中，酸奶较少被关注，由于其具有高酸度，并且牛奶的巴氏消毒被认为是控制包括单核细胞增生李斯特菌在内的许多病原菌的有效屏障。

4.原产于美国的甜柿子椒是茄属的水果，它的消费量呈流行趋势增长，主要是由于它具有丰富的颜色(范围由绿、黄、红到紫色)、形状、大小和特有的风味。柿子椒可以用来生产脱水产品(如辣椒粉)、腌辣椒、冻辣椒片或丁，可以用来做比萨饼或者沙拉。去年对于冷冻的辣椒片或者辣椒丁的需求大大地增加了，是因为消费者想要吃到只经过最低限度加工的蔬菜产品，并以此作为健康的饮食习惯的一部分。

5.生物可降解膜通常都是用生物高分子(如碳水化合物、蛋白质和脂肪)制成的。由碳水化合物制成的膜具有良好的机械特性，它们也是抵抗低极性成分的有效屏障，但是这些涂膜不能提供好的对抗湿度的屏障。另外，蛋白质膜对氧气、二氧化碳和一些芳香物质都是非常好的屏障，但是它们的机械特性并不令人满意，这些特性限制了它们在不同方面的应用。

6.奶酪是一种发酵乳制品。制作奶酪的原理是凝结乳以形成凝块和乳清。今天帮助形成凝乳的方法是添加前体细菌和粗制凝乳酶，该酶可以加速凝块和乳清的分离。通常，乳酪的制作是以酸化开始的，这是一个降低乳的 pH 的过程，通常这个过程是由细菌来完成的，它们以乳中的乳糖为食物，而产生废物乳酸。

7.现在，许多食品公司都提供即食的马铃薯制品(如煎薯条、薯角或者薯片)，它们是全世界都广泛消费的食品。这些制品在短时间地煎、微波加热、煮制或是直接用开水冲泡后即可食用。经过了这些简单的烹调的(马铃薯的)植物组织经历了一些理化变化，这些变化对最终产品的质地形成和质地特点都起了决定性的作用。这些质地(如结构、厚度、坚硬度、易碎性、颜色和味道等)对产品质量和消费者选择的影响都是非常重要的。

8.近年来关于果蔬的总抗氧化能力的研究,揭示了一大批有色化合物也许对自由基有抗氧化的保护作用,这些化合物中有一些是类黄酮(包括花青素、黄酮、异黄酮、儿茶酚和表儿茶酚)。另外,果蔬中还含有维生素和矿物质,植物化合物如类黄酮和其他多酚类物质也许可以通过给人体提供更强的抗氧化剂保护,降低发生心血管疾病和各种癌症的危险。

9.牛奶能经发酵生成各种在外观、香味、结构和健康性等特点上都不同的产品。生产这些不同产品的艺术在于不同的工艺和发酵工程中乳酸菌的选择。新技术如分子生物学和基因改良技术的出现,一方面增加了筛选成千上万种自然菌株和突变菌株的可能性,另一方面可以通过直接的基因工程改善菌株。本文列举了 2 个研发乳品发酵中前体菌株的例子。而且,按基因修饰的特点分成了 3 个组:①一步基因改变,如删除、基因扩大、质粒插入或缺失。②多步的同一种属的基因 DNA 重排。③跨种属间的基因修饰。在不同的国家,基因改变的特点对安全评估和立法含义都有影响。

10.食品安全中心是国家的一个非营利性组织,旨在通过促进有机农业和其他可持续性的实践活动来保护人类健康和环境。该组织从事立法提案动员大众和设计教育项目等活动,通过这些活动来影响政府和企业,告知公众诸如基因工程、食品放射性以及有机食品的标准等议题。

11.葡萄酒是一种复杂的化学液体,科学家们近些年来才开始了解它的特性。据估计葡萄酒含有超过 1 000 种挥发性的风味成分,其中超过 400 种是由酵母产生的。尽管已经研究了数十年,科学家们也只是现在才开始对葡萄酒的化学成分的特点有了一个全面的了解。部分困难是因为这是一个动力学的问题,各种风味成分的挥发性可以被葡萄酒中的其他成分改变。

12.依照各个国家饮食模式和烹调方法的不同,含有最高丙烯酰胺摄入量的食品也有所不同。一般来说,马铃薯制品、咖啡、焙烤食品是最重要的丙烯酰胺的来源。例如在德国,人们每天要吃掉大约 240 g 的面包和面包圈,使得丙烯酰胺摄入量的 25%都来源于它们。

第9章

参考文献

References

第1课 概 述

文后参考文献(bibliographic references)是指为撰写或编辑论文和著作而引用的有关文献信息资源。其主要作用是介绍前人的工作，提供考据资料；表明作者的水平和研究的起始点；反映作者的科学态度和对知识产权的尊重；反映论文的水平，便于读者查阅。

一、著录项目

(一)主要责任者或其他责任者

个人著者采用姓在前名在后的著录形式。欧美著者的名可以用缩写字母，缩写名后省略缩写点。欧美著者的中译名可以只著录其姓，同姓不同名的欧美著者，其中译名不仅要著录其姓，还需著录其名。用汉语拼音书写的中国著者姓名不得缩写。

著作方式相同的责任者不超过3个时，全部照录。超过3个时，只著录前3个责任者，其后加“，等”或与之相应的词。

无责任者或者责任者情况不明的文献，“主要责任者”项应注明“佚名”或与之相应的词。凡采用顺序编码制排列的参考文献可省略此项，直接著录题名。

凡是对文献负责的机关团体名称通常根据著录信息源著录。用拉丁文书写的机关团体名称应由上至下分级著录。

(二)题名

题名包括书名、刊名、报纸名、专利题名、科技报告名、标准文献名、学位论文名等。题名按著录信息源所载的内容著录。

(三)版本

第1版不著录，其他版本说明需著录。版本用阿拉伯数字、序数缩写形式或其他标志表示。

(四)出版项

出版项按出版地、出版者、出版年顺序著录。

1. 出版地

出版地著录出版者所在地的城市名称。对同名异地或不为人们熟悉的城市名，应在城市名后附省名、州名或国名等限定语。文献中载有多个出版地，只著录第一个或处于显要位置的出版地。

2. 出版者

出版者可以按著录信息源所载的形式著录，也可以按国际公认的简化形式或缩写形式著录。著录信息源载有多个出版者，只著录第一个或处于显要位置的出版者。

3. 出版日期

出版年采用公元纪年，并用阿拉伯数字著录。如有其他纪年形式时，将原有的纪年形式置于“(　)”内，报纸和专利文献需详细著录出版日期，其形式为“YYYY-MM-DD”。

(五)页码

专著或期刊中析出文献的页码或引文页码,要求用阿拉伯数字著录。

二、著录格式

(一)专著著录格式

主要责任者.题名:其他题名信息[文献类型标识].其他责任者.版本项.出版地:出版者,出版年:引文页码[引用日期].获取和访问路径.

举例:

1. Heldman D R, Moraru C I. Encyclipedia of Agricultural, Food, and Biological Engineering. 2nd ed. London: Taylor & Francis, 2010.

2. Rahman M S, Perera C. Drying and Food Preservation. In Handbook of Food Preservation; Rahman M S, ed.; NY: Marcel Dekker, Inc. 1999: 192-194.

(二)连续出版物著录格式

主要责任者.题名:其他题名信息[文献类型标识].年,卷(期)-年,卷(期). 出版地:出版者,出版年[引用日期].获取和访问路径.

举例:

1. Charanjit K, Binoy G. Viscosity and quality of tomato juice as affected by processing methods. Journal of Food Quality, 2007,30(6): 864-877.

2. Kerley C P, Elnazir B, Greally P, et al. Blunted serum 25(OH)D response to vitamin D_3 supplementation in children with autism. Nutritional Neuro Science, 2020, 23(7):537-542.

(三)专利文献著录格式

专利申请者或所有者.专利题名:专利国别,专利号[文献类型标识].公告日期或公开日期[引用日期].获取和访问路径.

举例:

Tachibana R, Shimizu S, Kobayshi S, et al. Electronic watermarking method and system: US6915001[P/OL]. 2005-07-05[2013-11-11].

(四)电子文献著录格式

主要责任者.题名:其他题名信息[文献类型标识/文献载体标识].出版地:出版者,出版年:引文(更新或修改日期)[引用日期].获取和访问路径.

举例:

Online Computer Library Center, Inc. About OCLC: history of cooperation [EB/OL]. [2012-03-27]. http://www. oclc. org / about /cooperation. en. html.

三、参考文献的标注体系

参考文献的标注体系有3种:①顺序编码制(numeric references method):按引文出现的先后标注数字,置于著者或叙述文字后的右上角,文后按第一次出现的次序排列。全世界各国包括中国采用此标注体系。②著者-出版年制(first element and date method):引文采用著者-出版年标注,参考文献表按著者字顺和出版年排序。美国的出版物常用著者-出版年制。③混合体系(CBE):文内为著者,但数字编码不按出现的次序排列,而按文后著者出版年体系编排的数序,文后按著者出版年编排,前面加序号。此处仅介绍顺序编码制和著者-出版年制。具体采用哪种标注格式,可查阅相关杂志的投稿须知。

(一)顺序编码制

顺序编码制是按正文中引用的文献出现的先后顺序连续编码,并将序号置于方括号中。同一处引用多篇文献时,只需将各篇文献的序号在方括号内全部列出,各序号间用“,”。如遇连续序号,可标注起讫序号。多次引用同一著者的同一文献时,在正文中标注首次引用的文献序号,并在序号的“[ ]”外著录引文页码。

例 1　*Journal of Agricultural and Food Chemistry*(美国)的参考文献标注格式。

Starch with a Slow Digestion Property Produced by Altering Its Chain Length, Branch Density, and Crystalline Structure

选自: J. Agric. Food Chem., 2007,55(11):4540-4547

作 者: Zihua Ao, Senay Simsek, Genyi Zhang, Mahesh Venkatachalam, Bradley L. Reuhs, Bruce R. Hamaker

Starch is the main glycemic carbohydrate in starchy foods. According to the rate and extent of starch digestion in vitro, starch has been classified into three major fractions [1] : (1) Rapidly digestible starch, the portion of starch digested within the first 20 min of incubation, (2) Slowly digestible starch, the portion of starch digested from 20 to 120 min, and,(3) Resistant starch, the remaining portion that cannot be further digested. A highly significant positive correlation between the glycemic index (GI) and rapidly digestible starch was reported [2]. GI is defined as the incremental area under the glucose response curve after a standard amount of carbohydrate from a test food is consumed relative to that of a control food (glucose or white bread)[3]. The human physiological consequences of GI have been related to diabetes, prediabetes, cardiovascular disease, and obesity [4]. High GI meals promoted fat deposition in mice, resulting in almost twice the body fat of those consuming low GI meals [5]. In obese teenagers, the rapid absorption of glucose after consumption of high GI meals induced a sequence of hormonal and metabolic changes that were related to excessive food intake [6]. Low GI meals decreased non fasting plasma glucose, plasma triacylglycerols, and adipocyte volume in rats [7] and prolonged satiety in obese adolescents [8] . Starch with a slow digestion property would provide for extended glucose (energy) release along with a low glycemic response and, thus, may have commercial application as a healthy ingredient of processed foods. There are no commercial slowly digestible starch-based products available in the current food market to our knowledge.

Reference

1. Englyst H N, Kingman S M, Cummings J H. Classification and measurement of

nutritionally important starch fractions. *Eur. J. Clin. Nutr*. 1992, 46, S33-S50.

2. Englyst H N, Veenstra J, Hudson G J. Measurement of rapidly available glucose (RAG) in plant foods: a potential in vitro predictor of the glycaemic response. *Br. J. Nutr*. 1996, 75, 327-337.

3. Wolever T M, Jenkins D A, Jenkins A L, Josse R G. The glycemic index: methodology and clinical implications. *Am. J. Clin. Nutr*. 1991, 54, 846-854.

4. Ludwig D S. The glycemic index: physiological mechanisms relating to obesity, diabetes, and cardiovascular disease. *J. Am. Med. Assoc*. 2002, 287, 2414-2423.

5. Pawlak D B, Kushner J A, Ludwig D S. Effects of a dietary glycaemic index on adiposity, glucose homeostasis, and plasma lipids in animals. *Lancet*. 2004, 364, 778-785.

6. Ludwig D S, Majzoub J A, Al-Zahrani A, Dallal G E, Blanco I, Roberts S B. High glycemic index foods, overeating, and obesity. *Pediatrics*. 1999, 103, E26.

7. Lerer-Metzger M, Rizkalla S W, Luo J, Champ M, Kabir M, Bruzzo F, Bornet F, Slama G. Effects of long-term low-glycaemic index starchy food on plasma glucose and lipid concentrations and adipose tissue cellularity in normal and diabetic rats. *Br. J. Nutr*. 1996; 75, 723-732.

8. Ball S D, Keller K R, Moyer-Mileur L J, Ding Y W, Donaldson D, Jackson W D. Prolongation of satiety after low versus moderately high glycemic index meals in obese adolescents. *Pediatrics*. 2003, 111, 488-494.

例 2 *Journal of Nutrition*(美国)的参考文献标注格式

Carob Pulp Preparation Rich in Insoluble Dietary Fiber and Polyphenols Enhances Lipid Oxidation and Lowers Postprandial Acylated Ghrelin in Humans

选自: J. Nutr. ,2006, 136:1533-1538

作者: Sindy Gruendel, Ada L Garcia, Baerbel Otto, Corinna Mueller, Jochen Steiniger, Martin O Weickert, Maria Speth, Norbert Katz, Corinna Koebnick

Dietary fiber consumption has been suggested to influence hunger, satiety, and energy intake in humans [1,2]. Dietary fiber is a complex group of substances, commonly divided into soluble and insoluble fibers. Soluble fiber prolongs gastric emptying and macronutrient absorption, which is linked to delayed hunger feelings and decreased energy intake [1,3]. Soluble fiber was shown to improve glucose homeostasis and lipid profile [4,5] and has been associated with short-term reduced energy intake in obese adults[3]. Insoluble dietary fiber increased the rate of glucose disappearance [6] and improved carbohydrate handling [7]. The effects of dietary fiber consumption on body weight management may be related to gut hormones, which regulate satiety and energy intake [1]. Ghrelin is a peptide hormone produced and excreted mainly in the stomach [8], and circulating in two major forms, acylated ghrelin and desacyl ghrelin[9,10]. Acylated ghrelin acts as an orexigenic signal to the central nervous system, with increased levels during fasting, and suppressed levels postprandially [11,12]. The administration of ghrelin induces body weight gain in rodents by promoting food intake and decreasing fat utilization [8,13]. In rodents, ghrelin administration increased the respiratory quotient (RQ), suggesting decreased fatty acid oxidation and increased glycolysis [11,13]. Carob pulp preparation (carob fiber), derived from the bean-like fruits of *Ceratonia siliqua*, is rich in insoluble dietary fiber and polyphenols. In humans, consumption of carob fiber was shown to have a high antioxidant capacity [14] and to lower serum cholesterol and serum triglycerides [15]. Furthermore, other studies showed that polyphenols may increase fat oxidation and energy expenditure in humans [16] and in mice [17]. Therefore, carob fiber may exert beneficial effects on postprandial lipid metabolism and substrate utilization potentially related to the secretion of gut hormones.

LITERATURE CITED

1. Howarth N C, Saltzman E, Roberts S B. Dietary fiber and weight regulation. Nutr Rev. 2001;59:129-139.

2. Slavin J L. Dietary fiber and body weight. Nutrition. 2005,21:411-418.

3. Pasman W J, Saris W H, Wauters M A, Westerterp-Plantenga M S. Effect of one week of fibre supplementation on hunger and satiety ratings and energy intake. Appetite. 1997,29:77-87.

4. Moreno L A, Tresaco B, Bueno G, Fleta J, Rodriguez G, Garagorri J M, Bueno M. Psyllium fibre and the metabolic control of obese children and adolescents. J Physiol Biochem. 2003,59:235-242.

5. Jenkins D J, Wolever T M, Leeds A R, Gassull M A, Haisman P, Dilawari J, Goff D V, Metz G L, Alberti K G. Dietary fibres, fibre analogues, and glucose tolerance: importance of viscosity. BMJ. 1978,1:1392-1394.

6. Schenk S, Davidson C J, Zderic T W, Byerley L O, Coyle E F. Different glycemic indexes of breakfast cereals are not due to glucose entry into blood but to glucose removal by tissue. Am J Clin Nutr. 2003,78:742-748.

7. Weickert M O, Mohlig M, Koebnick C, Holst J J, Namsolleck P, Ristow M, Osterhoff M, Rochlitz H, Rudovich N, et al. Impact of cereal fibre on glucose-regulating factors. Diabetologia. 2005,48:2343-2353.

8. Kojima M, Kangawa K G, an orexigenic signaling molecule from the gastrointestinal tract. Curr Opin Pharmacol. 2002,2:665-668.

9. Kojima M, Hosoda H, Date Y, Nakazato M, Matsuo H, Kangawa K. Ghrelin is a growth-hormone-releasing acylated peptide from stomach. Nature. 1999:402:656-60.

10. Korbonits M, Goldstone A P, Gueorguiev M, Grossman A B. Ghrelin-a hormone with multiple functions. Front Neuroendocrinol. 2004,25:27-68.

11. Wren A M, Seal L J, Cohen M A, Brynes A E, Frost G S, Murphy K G, Dhillo

W S, Ghatei M A, Bloom S R. Ghrelin enhances appetite and increases food intake in humans. J Clin Endocrinol Metab. 2001,86:5992-5995.

12. Cummings D E, Purnell J Q, Frayo R S, Schmidova K, Wisse B E, Weigle D S. A preprandial rise in plasma ghrelin levels suggests a role in meal initiation in humans. Diabetes. 2001,50:1714-1719.

13. Tschop M, Smiley D L, Heiman M L. Ghrelin induces adiposity in rodents. Nature. 2000,407:908-913.

14. Kumazawa S, Taniguchi M, Suzuki Y, Shimura M, Kwon M S, Nakayama T. Antioxidant activity of polyphenols in carob pods. J Agric Food Chem. 2002,50:373-377.

15. Zunft H J, Luder W, Harde A, Haber B, Graubaum H J, Koebnick C, Grunwald J. Carob pulp preparation rich in insoluble fibre lowers total and LDL cholesterol in hypercholesterolemic patients. Eur J Nutr. 2003,42:235-242.

16. Dulloo A G, Duret C, Rohrer D, Girardier L, Mensi N, Fathi M, Chantre P, Vandermander J. Efficacy of a green tea extract rich in catechin polyphenols and caffeine in increasing 24-h energy expenditure and fat oxidation in humans. Am J Clin Nutr. 1999, 70:1040-1045.

17. Klaus S, Pultz S, Thone-Reineke C, Wolfram S. Epigallocatechin gallate attenuates diet-induced obesity in mice by decreasing energy absorption and increasing fat oxidation. Int J Obes (Lond). 2005,29:615-623.

例 3 *Carbohydrate Research*(荷兰)的参考文献标注格式

Water-soluble 3-*O*-(2-methoxyethyl)cellulose: synthesis and characterization

选自：Carbohydrate Research，2008，343(4):668-673

作者：Thomas Heinze，Andreas Koschella

Cellulose ethers are important biopolymer derivatives used in various applications. In particular, mixed cellulose ethers including hydroxyethyl- and hydroxypropylmethylcelluloses are commercially produced in large scale and are applied in various fields.[1] The technical synthesis is carried out by conversion of alkali cellulose with the corresponding epoxide and methyl chloride to give a random distribution of the substituents within the anhydroglucose unit and along the polymer chain. Moreover, the newly formed hydroxyl groups of the hydroxyethyl- and hydroxypropyl moieties may be methylated. It is well known that the distribution of functional groups may influence the properties of cellulose ethers.[2] To gain detailed information about the influence of the structure on properties, and hence to improve the quality of the technical cellulose ethers, not only a comprehensive structure characterization but also cellulose ethers with a defined functionalization pattern are indispensable for the establishment of the structure-property relationships. Regarding commercially important cellulose ethers, 3-*O*-methylcellulose synthesized via 2,6-di-*O*-thexyldimethylsilylcellulose is insoluble in water and organic liquids,[3] while 3-*O*-ethylcellulose is well soluble in water showing a different thermal transition temperature compared to conventional ethyl cellulose.[4] These results lead to the question of whether 3-*O*-(2-methoxyethylcellulose)(MEC) can be synthesized via 2,6-di-*O*-protected cellulose. As already mentioned, a 2-methoxyethyl moiety may be present in commercial hydroxyethylcellulose due to methylation of the hydroxyl group formed by the reaction of the ethylene oxide with the biopolymer. Moreover, the properties of regioselectively functionalized 3-*O*-MEC are of interest. In the present paper, the synthesis and detailed structure characterization of 3-*O*-MEC is discussed.

References

1. Brandt L. Cellulose Ethers. In *Ullmann's Encyclopedia of Industrial Chemistry*; Gerhartz W, Yamamoto Y S, Campbell F T, Pfefferkorn R, Rounsaville J F, Eds.; VCH: Weinheim, 1986; Vol. A5, pp 461-488.

2. Heinze T. Chemical Functionalization of Cellulose. In *Polysaccharides: Structural Diversity and Functional Versatility*; Dumitriu S, Ed.; Marcel Dekker: New York, 2004; pp 551-590.

3. Koschella A, Heinze T, Klemm D. *Macromol. Biosci*. 2001, 1, 49-54.

4. Koschella A, Fenn D, Heinze T. *Polym. Bull*. 2006, 57, 33-41.

(二)著者-出版年制

正文引用的文献采用著者-出版年制时,各篇文献的标注内容由著者姓氏与出版年构成,并置于“(　)”内。倘若只标注著者姓氏无法识别该人名时,可标注著者姓名,集体著者著述的文献可标注机关团体名称。在正文中引用多著者文献时,对欧美著者只需标注第一个著者的姓,其后附“et al”;正文中引用同一著者相同年号发表的不同文献时,按正文中出现的先后顺序在年号后用小写字母区分,在参考文献表中以同样方法加以区分。多次引用同一著者的同一文献,在正文中标注著者与出版年,并在“(　)”外以角标的形式著录引文页码。参考文献列表时则按作者姓氏的第一字母排序。

例 1　*Food Chemistry*(美国)参考文献标注格式

Extractable oil in microcapsules prepared by spray-drying: Localisation, determination and impact on oxidative stability

选自: Food Chemistry, 2008, 109(1): 17-24

作者: S. Drusch and S. Berg

Microencapsulation techniques offer the possibility for the protection and controlled release of food ingredients. Generally, encapsulation techniques can be divided into three classes: chemical processes like molecular inclusion or interfacial polymerization, physico-chemical techniques like coacervation and liposome encapsulation and physical processes like spray-drying, spray chilling/cooling, co-crystallization, extrusion or fluidized bed coating (Kunz, Krückeberg, & Weiβbrodt, 2003). The principal technologies used for encapsulation of lipophilic food ingredients are spray-drying, coacervation and extrusion.

In spray-dried emulsions, the amount of non-encapsulated oil is a key parameter determining the product quality. It has been shown, that in fat-containing dairy powders the surface is almost completely covered with a thin layer of fat (Kim, Chen, & Pearce, 2002) and it is a well-established fact that this surface fat determines the flowability and wettability of spray-dried dairy powders (Vega & Roos, 2006). Furthermore, non-encapsulated milk fat undergoes oxidation and may lead to off-flavour formation. It has already been shown in the early 1970s for spray-dried dairy products by Buma (1971a), that the extractable oil consists of different fractions comprising (1) the surface fat, (2) the outer layer fat in the surface layer of the particle, (3) fat, that can be extracted by the solvent through capillary forces and (4) fat, that can be reached by solvent through holes left by already extracted fat (Fig. 1). Several different methods for the determination of free or surface fat in spray-dried emulsions exist (Vega & Roos, 2006) and these different methods extract different fractions of the non-encapsulated oil.

References

Buma T J. (1971a). Free fat in spray-dried whole milk. 1. General introduction and brief review of literature. *Netherlands Milk and Dairy Journal*, 25, 33-41.

Kim E H J, Chen X D, & Pearce D. (2002). Surface characterization of four industrial spray-dried dairy powders in relation to chemical composition, structure and wetting property. *Colloids and Surfaces B: Biointerfaces*, 26, 197-212.

Minemoto Y, Adachi S, & Matsuno R. (1997). Comparison of oxidation of methyl linoleate encapsulated with gum arabic by hot-airdrying and freeze-drying. *Journal of Agricultural and Food Chemistry*, 45, 4530-4534.

Vega C, & Roos Y H. (2006). Spray-dried dairy and dairy-like emulsions-compositional considerations. *Journal of Dairy Science*, 89, 383-401.

例 2 *Trends in Food Science & Technology*(美国)的参考文献标注格式

Understanding and modelling bacterial transfer to foods: a review

选自: Trends in Food Science & Technology, 2008, 19(3): 131-144

作者: F. Pérez-Rodríguez, A. Valero, E. Carrasco, R. M. García and G. Zurera

The well-established UK surveillance system reported that cross contamination was the main contributing factor (32%) in the outbreaks investigated in the period 1999-2000 (WHO, 2003). Similarly, the US Centers for Disease Control and Prevention (CDC) reported that 18% and 19% of food-borne diseases caused by bacteria in the years 1993-1997 in the United States were associated with contaminated equipment and poor hygiene practices, respectively (CDC, 2000). Moreover, although most outbreaks result from extensive growth at abusive storage temperatures, insufficient cooking, etc., many are also associated with bacterial cross contamination/recontamination (Roberts, 1990). Similarly, various authors have stated that cross contamination of bacterial and viral pathogens in homes and in food-service establishments could well be the major contributing factor to sporadic and epidemic food-borne illnesses ([Beumer and Kusumaningrum, 2003], [Bloomfield, 2003] and [Chen et al., 2001]). Therefore, it is apparent that pathogens transfer to foods through many different types of events such as recontamination and cross contamination, etc. might be decisive in many of the outbreaks studied.

References

Beumer R R, & Kusumaningrum H. (2003). Kitchen hygiene in daily life. *International Biodeterioration & Biodegradation*, 51, 299-302.

Bloomfield S F. (2003). Home hygiene: a risk approach. *International Journal of Hygiene and Environmental Health*, 206, 1-8.

Bremer P J, Fillery S, & McQuillan A J. (2006). Laboratory scale Clean-In-Place (CIP) studies on the effectiveness of different caustic and acid wash steps on the removal of dairy biofilms. *International Journal of Food Microbiology*, 106, 254-262.

Chen Y, Jackson K M, Chea F P, & Schaffner D W. (2001). Quantification and variability analysis of bacterial cross-contamination rates in common food service tasks. *Journal of Food Protection*, 64, 72-80.

Roberts D. (1990). Foodborne illness, sources of infection: food. *Lancet*, 336, 859-861.

WHO (World Health Organization) (2003). Eighth report 1999-2000 of WHO Surveillance Programme for control of foodborne infections and intoxications in Europe. Available from http://www.bfr.bund.de/internet/8threport/8threp_fr.htm. Accessed 14.10.06.

第2课　实　　例

一、著作

1. Lee F A. The Blanching Process. In Advances in Food Research. NY: Academic 1958: 63-109.

2. Rahman M S, Perera C. Drying and Food Preservation. In Handbook of Food Preservation. Rahman M S, Ed. NY: Marcel Dekker, Inc., 1999: 192-194.

3. John M. Wilkins Shaun Hill. Food in the Ancient World. Ancient Cultures. Malden, Mass, 2006.

4. Mead G C. Food safety control in the poultry industry. Woodhead Publishing, 2005.

5. Ronald S J. Wine science. 3rd ed. Academic press, 2008.

6. Hui Y H. Handbook of Food and Beverage Fermentation Technology. Marcel Dekker, Inc., 2004.

7. James N, Parkr M D. Food additives. Woodhead Publishing, 2004.

8. Coles R. Dowell M D, Kirwan J M. Food Packaging Technology. Blackwell Publishing, 2003.

9. Zacharias B M, George D S. Food Process Design. Marcel Dekker, Inc., 2003.

10. Leveld H L, Mostert M A, Holah J and White B. Hygiene in Food Processing. Woodhead Publishing, 2003.

11. Zeuthen P and Bogh-Sorensen L. Food preservation techniques. Woodhead Publishing, 2003.

12. Taylor A J. Food Flavour Technology. Blackwell Publishing, 2002.

13. Blanchfield J R. Food labeling. Woodhead Publishing, 2002.

14. Friberg S E. Food Emulsions. 4th ed. Revised and Expanded. Marcel Dekker, Inc., 2004.

15. Lelieveld H L M, Mostert M A and Holah J. Handbook of hygiene control in the food industry. Woodhead Publishing, 2005.

16. Wood R, Foster L, Damant A and Key P. Analytical methods for food additives. Woodhead Publishing, 2004.

17. Arora D K. Fungal Biotechnology in Agricultural, Food, and Environmental Applications. Marcel Dekker, 2004.

18. Whitehurst J R. Emulsifiers in food technology. Macsource press, 2004.

19. Bernard D, Lockwood A, Alcott D and Pantelids I. Food and beverage management. Elsevier Inc., 2012.

20. Hall C M, Sharples L, Mitchell R, Macionis N, Cambourne B. Foodtourism around the world. Butterworth-Heinemann, 2003.

二、期刊

1. Stanley D W, Bourne M C, Stone A P, et al. Low temperature blanching effects on chemistry, firmness and structure of canned green beans and carrots. Food Sci, 1995, 60:327-333.

2. Lin Z, Schyvens E. Influence of blanching treatments on the texture and color of

some processed vegetables and fruits. Food Process, 1995,19:451-465.

3. Mechteldis G E W. Prediction of degrade-ability of starch by gelatinization menthalpy as measure by differential scanning calorimetry. Starch,1992(1):14-18.

4. Jane J, Shen L, Wang L, et al. Preparation and properties of small-praticle corn starch. Cereal Chemistry,1992,69(3):280-283.

5. Balachandran S, Kentish S E, Mawson R, et al. Ultrasonic enhancement of the supercritical extraction from ginger. Ultrasonic Sonochemistry,2006,13(6):471-479.

6. Kimbaris A C, Siatis N G, Daferera D J, et al. Comparison of distillation and ultrasound-assisted extraction methods for the isolation of sensitive aroma compounds from garlic. Ultrasonics Sonochemistry,2006,13(1):54-60.

7. To E C, Mudgett R E, Wang D I C, et al. Dielectric properties of food materials. Food Science and Nutrition, 2004,44:465-471.

8. Kraszewski A W, Nelson S O. Microwave permittivity determination in agricultural products. Journal of microwave power and electromagnetic energy, 2004, 39 (1):41-52.

9. Mao W J, Watanabe M and Sakai N. Dielectric properties of surimi at 915 MHz as affected by temperature, salt and starch. Fisheries science, 2003,69:1042-1047.

10. Mateo A, Soto F, Villarejo J A. Quality analysis of tuna meat using an automated color inspection system. Agricultural Engineering,2006,35(1):1-13.

11. Lin C H, Chang C Y. Textural change and antioxidant properties of broccoli under different cooking treatments. Food chemistry,2005,90(1):9-15.

12. Singh G D, Sharma R, Bawa A S, et al. Drying and rehydration characteristics of water chestnut (Trapa natans) as a function of drying air temperature. Journal of Food Engineering, 2008,87(2):213-221.

13. Bozzoli F, Cattani L, Rainieri, et al. Estimation of Cocal heat transfer coefficient in coiled tubes under inverse heat conduction problem approach. Experimental Thermal and Fluid Science, 2014, 59:246-251.

14. Sacilik K and Elicin A K. The thin layer drying characteristics of organic apple slices. Journal of Food Engineering,2006,73(3): 281-289.

15. Jayaraman K S, Das Gupta D K, Babu Rao N. Effects of pre-treatment with salt and sucrose on quality and stability of dried cauliflower. Journal of Food Science, 1990,25: 47-51.

16. Torringo E, Esveld E, I Scheewe, et al. Osmotic dehydration as a pre-treatment before combined microwave-hot-air drying of mush-room. Journal of Food Engineering, 2001,49:185-191.

17. Raghav P K, Grover PC and Thapar V K. Storage stability of sunflower (*Helianthus annus*) oil in different packaging materials. Journal of Food Science, 1999,

36，253-255.

18. Charanjit K，Binoy G，Deepa N，et al. Viscosity and quality of tomatojuice as affect by processing methods. Journal of Food Quality,2007,30(6)：864-877.

19. Song H and Cadwallader K R. Aroma components of american country ham. Journal of Food Science，2008,73(1)：29-35.

20. Agüero M V，Barg M V，Yommi A，et al. Postharvest changes in water status and chlorophyll content of lettuce（*Lactuca Sativa* L.）and their relationship with overall visual quality. Journal of Food Science,2008,73(1)：47-55.

三、专利

1. Pacifico C J，Wu W H，Fraley M. Sensitive substance encapsulation：US09996636. 2001-11-29.

2. Coughlin M F，Cole D C，Crawford C A. Method for cleaning and/or disinfecting food processing equipment：US10898093. 2008-02-12.

3. Brundage T J，Rodriguez T F. Digital watermarking apparatus and methods：US11622202.2007-01-11.

4. Koka R，Mehnert D W，Fritsch R J. Process for manufacturing cheeses and other dairy products and products thereof：US11038355.2005-01-19.

5. Alexander J G，Prashant G，Vaibhav G，et al. Food transportation system：US 201916515911. 2019-07-18.

6. Kawamura M. Food preserving method and its device：US13851060. 2013-03-26.

7. Petrella J A. Process for Making Jelly Containing Pectin：US20060462752. 2006-08-07.

8. Pogosjan A S. Method for producing special beer：RU20070108879.2007-03-12.

9. Vanhemelrijck J G，Mccrae C H. Citrus fruit fibres in emulsions：AU2006265336. 2006-07-03.

10. Strohm A G，Anderson R K. Produce a preservation system：US20050095633. 2005-03-30.

四、电子出版物

1. Van Alfen N K. Encyclopedia of Agriculture and Food Systems,Academic Press, 2014.[2018-06-25].

https：// www. sciencedirect. com/referencework/9780080931395/encyclopedia-of-agriculture-and-food-systems＃book-info

2. Colgrave M L. Proteomics in Food Science. Academic Press,2017.[2019-04-20].

https：// www. sciencedirect. com/book/9780128040072/proteomics-in-food-science＃book-info

3. Galanakis C M. Gastronomy and Food Science. Academic Press，2020.[2020-10-25].

https：// www.sciencedirect.com/book/9780128200575/gastronomy-and-food-science

4. Busquets R. Emerging Nanotechnologies in Food Science. Elsevier, 2017.［2019-09-25］.

https://www.sciencedirect.com/book/9780323429801/emerging-nanotechnologies-in-food-science#book-info

5. Ashutosh K S. Electron Spin Resonance in Food Science. Academic Press, 2017.［2018-10-21］.

https://www.sciencedirect.com/book/9780128054284/electron-spin-resonance-in-food-science

6. Andrei A B, Hassan Y A, Vu D H. Vibrational Spectroscopy Applications in Biomedical, Pharmaceutical and Food Sciences. Elsevier, 2020.［2021-04-05］.

https://www.sciencedirect.com/book/9780128188279/vibrational-spectroscopy-applications-in-biomedical-pharmaceutical-and-food-sciences

7. Neil S I. High Pressure Food Science, Bioscience and Chemistry. Woodhead Publishing, 1998.［2015-09-25］.

https://www.sciencedirect.com/book/9781855738232/high-pressure-food-science-bioscience-and-chemistry

8. Whistler R L, Bemiller J N, Paschall E F. Starch: Chemistry and Technology (Second Edition). Academic Press, 1984.［2001-09-25］

https://www.sciencedirect.com/book/9780127462707/starch-chemistry-and-technology

9. Batt C A, Tortorello M L. Encyclopedia of Food Microbiology (Second Edition). Academic Press, 2014.［2019-10-15］.

https://www.sciencedirect.com/referencework/9780123847331/encyclopedia-of-food-microbiology

10. Membré J M, Valdramidis V. Modeling in Food Microbiology. ISTE Press-Elsevier, 2016.［2019-04-05］.

https://www.sciencedirect.com/book/9781785481550/modeling-in-food-microbiology

11. Abbas F M A, Wasin A A A. Easy Statistics for Food Science with R. Academic Press, 2019.［2021-03-12］.

https://www.sciencedirect.com/book/9780128142622/easy-statistics-for-food-science-with-r

12. James J M, Burks W, Eigenmann P. Food Allergy. Saunders, 2012.［2015-09-25］.

https://www.sciencedirect.com/book/9781437719925/food-allergy

13. Kunal P, Indranil B, Preetam S, Arindam Bit, Doman Kim, Arfat Anis, Samarendra Maji. Food, Medical, and Environmental Applications of Polysaccharides. Elsevier, 2021.［2021-04-20］.

https://www.sciencedirect.com/book/9780128192399/food-medical-and-environmental-applications-of-polysaccharides

14. Holban A M, Grumezescu A M. Foodborne Diseases. Academic Press, 2018. [2019-08-25].

https://www.sciencedirect.com/book/9780128114445/foodborne-diseases

15. Barbosa-Cánovas G, Mortimer A, Lineback D, Spiess W, Buckle K, Colonna P. Global Issues in Food Science and Technology. Academic Press, 2009. [2016-07-05].

https://www.sciencedirect.com/book/9780123741240/global-issues-in-food-science-and-technology

16. Kobun R. Advanced Food Analysis Tools. Academic Press, 2020. [2021-03-02].

https://www.sciencedirect.com/book/9780128205914/advanced-food-analysis-tools

17. Scott A O. Biosensors for Food Analysis. Woodhead Publishing, 1998. [2010-04-20].

https://www.sciencedirect.com/book/9781855737761/biosensors-for-food-analysis

18. Galanakis C M. Innovative Food Analysis. Academic Press, 2021. [2021-04-21].

https://www.sciencedirect.com/book/9780128194935/innovative-food-analysis

19. Fellows PJ. Food Processing Technology(Fourth Edition). Woodhead Publishing, 2017. [2018-04-25].

https://www.sciencedirect.com/book/9780081019078/food-processing-technology

20. Grumezescu A M, Holban A M. Food Processing for Increased Quality and Consumption. Academic Press, 2018. [2020-06-25].

https://www.sciencedirect.com/book/9780128114476/food-processing-for-increased-quality-and-consumption

21. Ades G, Leith K, Leith P. Food Safety. Academic Press, 2016. [2020-05-25].

https://www.sciencedirect.com/book/9780128031049/food-safety

22. Mead G C. Food Safety Control in the Poultry Industry. Woodhead Publishing, 2005. [2015-12-25].

https://www.sciencedirect.com/book/9781855739543/food-safety-control-in-the-poultry-industry

作 业 汉译英

1. 汉斯·黎曼，迪安·克里夫. 食源性传染病和中毒. 国际标准图书编号：978-0-12-588365-8. 出版时间:2005.

2. 营养丰富的紧急救援食品. 华盛顿特区:美国科学院医学研究所出版社.

3. 赵思明,俞兰苓,熊善柏,等. 稻米支链淀粉的流变学特性. 农业机械学报,2003(3):58-60.

4. 罗伯特 J. 怀特·赫斯特. 食品技术中的乳化剂.

5. 马丁 A. 阿伯拉罕. 可持续性科学与工程.

6. 界定原则. 国际标准图书编号：978-0-444-51712-8. 出版时间：2006.

7. 多尔食品公司专家. 食品百科全书. 健康营养指南. 国际标准图书编号：978-0-12-219803-8. 出版日期：2002.

8. 乔纳森·布罗斯托夫，斯蒂芬·夏拉柯姆. 食物过敏和排斥. 国际标准图书编号：978-0-7020-2038-4. 出版时间：2002.

9. 消费者食品安全和食品相关研究的国际宣传期刊：英国食品日报，0007-070X ；V. 107，No. 7，2005.1.

10. 罗登林，聂英，钟先锋. 超声强化超临界 CO_2 萃取人参皂苷的研究，农业工程学报，2007，6：256-258.

11. 许克勇，吴彩娥，李元瑞，等. 超临界二氧化碳萃取蒜汁中大蒜油的研究. 农业工程学报，2005，21(4)：150-154.

12. 罗登林，丘泰球，王睿瑞. 超声对超临界 CO_2 反相微乳萃取人参皂苷的影响. 江苏大学学报(自然科学版)，2006，27(3)：202-206.

13. 袁易全. 近代超声原理及应用. 南京：南京大学出版社，1996.

14. 廖传华，黄振仁. 超临界 CO_2 流体萃取技术、工艺开发及其应用. 北京：化学工业出版社. 2004：28-29.

15. 胡飞，陈玲，李琳，等. 微细化马铃薯淀粉流变学特性的研究. 中国粮油学报，2003，18(2)：61-63.

16. 胡飞，陈玲，李琳. 马铃薯淀粉颗粒在微细化过程中结晶结构的变化. 精细化工，2002，19(2)：114-117.

第 10 章

食品科技写作常用句型

Common Patterns of Food Scientific Thesis

第 1 课　摘要(Abstract)

英文摘要一般以第三人称、过去式描述。用词需准确，逻辑性强，结构严谨。结果和结论一般用句子表达，而目的、方法(设计、地点和对象)等，则常用短语表达。

常见句型如下：

(1)It was concluded that...

(2)The results illustrated/revealed that...

(3)A method was developed/designed/introduced/performed to/for...

(4)A simple method for... cases of... was described.

(5)The major goal of this investigation/the research was to evaluate...

(6)The studies/tests/researches (have) showed/suggested that...

(7)This investigation/research has been made to point out...

(8)The present study was designed with a view to clarify/determine/establish...

(9)This study intended to clarify/determine/examine/check/establish /demonstrate / provide evidence for/bring out...

(10)The main purpose of this study was to find out/establish/assess/analyze /understand...

(11)This paper was carried out/performed/attempted in order to demonstrate...

(12)... was analyzed/carried out/investigated/explored in this article.

第 2 课　引言(Introduction)

引言(也称前言、序言或概述)作为科技论文正文的第一部分，应以简短的篇幅介绍论文的写作背景和目的，以及相关领域前人所做的工作和研究的内容，说明本研究与前人工作的关系，目前研究的热点、存在的问题及本研究的意义，引出文章主题。

常见句型如下：

(1)Several studies have investigated/examined/explored/discussed/considered/ reported on...

(2)Some experiments about... have been published/done.

(3)Many studies/researches/investigations on... have been performed.

(4)A number of/Few researchers have studied...

(5)The term is considered to be/is taken to be/refers to...

(6)The word to be defined is a kind of class which distinguishing features...

(7)It should be mentioned that...

(8)In 2015,somebody showed/reported/observed/pointed out that...

(9)It was/has been investigated by... using... method.

(10)However, the work/research/attention concerning... is rare.

(11)However, few studies have been done/reported on...

(12)However, little researches have studied/have been devoted to...

(13)However, little attention has been paid to...

(14)However, little information has been published concerning...

(15)However, little is known about/available on...

(16)The experiment has been conducted in order to do.....

(17)This article aims at...

(18)The experiment was designed to/intend to...

第3课　材料与方法(Materials and Methods)

这一部分主要用于说明实验的对象、条件、使用的材料、实验步骤或计算过程等。该部分对过程的描述要完整具体,符合逻辑。相对于其他部分而言,材料与方法是英文科技论文中相对容易撰写的部分。

常见句型如下:

(1)To improve the efficiency of the method, the following approach was applied.

(2)The procedure we followed has certain advantages over the existing method...

(3)This models falls/can be divided/was classified into...

(4)There are two major groups/classes/categories/broad, including... and...

(5)The following steps should be followed.

(6)... was extracted/analyzed using the methods described by...

(7)The process has... major steps associated with it. First... second... third... The final step is...

第4课　结果与讨论(Results and Discussion)

对于一篇科技论文,"Results and Discussion"是核心部分,篇幅往往占到正文的50%以上。在结果和讨论不易明显分开的情况下,这两部分可以合起来撰写。

Results应直接描述本项研究所取得的所有结果,并对结果进行定量或定性的分析,揭示有关的数据和资料,简洁、清楚、直观地说明事实。对实验结果的表达要高度概括和提炼,不要简单地将实验记录数据或观察事实堆积到论文中,应突出具有科学意义和代表性的数据。所描述的实验结果要真实,即使有些结果与理论不符,也不可省略,并在讨论中加以说明和解释。

Discussion是对论文研究结果的分析和深化,在充分明确事实的基础上,旁征博引,从不同角度将本研究的结果与前人研究结果进行比较分析,指出与以往研究的异同,提出本项研究的创新性,阐述结果的意义,提出本项研究结果可能具有的广泛的科学意义、学术价值及应用前景。撰写要条理分明,逻辑清楚。

常见句型如下：

(1)These results suggested/recommended that...

(2)As shown in Table 1/Figure 1, we can find that...

(3)... are summarized/compared/listed in Table 1.

(4)These findings are understandable because...

(5)This circumstance may be caused by...

(6)These results agreed well with the findings of...

(7)These findings were in accordance with the data of...

(8)These results were highly consistent with the findings of...

(9)The data of our system was significantly higher than that of...

(10)The results can be explained by...

(11)A possible explanation for this is that...

(12)This may have occurred because...

(13)Definitions as given here apply only to the subjected treated in this standard.

(14)To some extent, these methods are only applicable to long series of production.

(15)This method is fitted to...

(16)The results obtained agreed with/fitted into...

(17)The result obtained was in agreement/line/consistent with...

(18)This can be illustrated by...

(19)This can be demonstrated the thought...

(20)A specific case can be provided to...

(21)An example of this involves...

(22)Another typical example is...

(23)The result can be categorized into... classes, including...

(24)From the results obtained so far, it seemed that...

(25)The experimental results for... are reported in Table 1.

(26)Therefore, it is reasonable to postulate that...

(27)Surprisingly, our result does not show the same trend reported in the previous references.

(28)This is in accordance with the results discussed in the previous study and supports that...

(29)The results supported/confirmed the assumption/hypothesis/evidence that

(30)This suggests/elucidates/means that...

(31)Therefore, we may conclude that...

(32)So, it is reasonable to postulate that...

(33)A careful study of... revealed that...

(34)The new findings from the experiment agree well with the results obtained in...

(35) Compared with the current research, the previous work was in connection with...

(36)An improvement on the result shown above can be made by based on the data provided.

(37)Based on the information contained in this research, we can conclude that...

(38)Although much effort has been made to..., this reality is far from completion.

(39)Because of the inaccuracy of the..., a conclusion cannot be drawn as...

(40)The results indicated that the total benefits were higher than the total costs.

(41)It should be noted that this study has examined only...

(42)We concentrate/focus on only...

(43)We have to point out that we do not...

(44)From the above discussion, the conclusion can be reached that...

第5课 结论(Conclusion)

英文食品科技论文一般都有结论,是在理论和实验的基础上,对结果进行逻辑推理后获得的总结性判断,与引言相互呼应。结论是本项研究的精华和浓缩表达。

常见句型如下:

(1)... has been studied in this article.

(2)Through..., it has demonstrated that...

(3)Overall, our study has revealed that...

(4)This study/investigation clearly demonstrated/discovered/represented/discussed...

(5)This research has shown/proposed/described...

(6)Overall, our study has revealed/shown...

(7)This research might provide some additional useful information to...

(8)From the results, we may conclude that...

(9)This paper has presented a theoretical and experimental study of...

(10)We have attempted to introduce some concepts associated with a theory of... based on...

(11)Considerable more work, hopefully, will be done in this area.

(12)Further study will be focused on...

(13)The problem requires further research/study/investigation/work in...

第6课 综述(Review)

与研究型科技论文相比,综述型科技论文主要是对某一科学课题或技术领域在一定时期发展状况的回顾总结,对现状的分析评论,对未来的预测展望。综述虽不一定具有首创性,但应有指导性,对科学技术的发展可起到承前启后的作用。其正文部分包括对所论述课题的概要介绍、该课题的进展及其研究价值及对已发表研究成果的比较讨论。

常用句型如下：

(1)In this article, a brief view of... is given.

(2)In the next section, a short review of... is given with special regard to...

(3)This paper will propose/present/discuss...

(4)The work of... made a major contribution to...

(5)The theory based on... is summarized/discussed in this article...

(6)Another major work on... by... further developed the concept of...

(7)The latest studies about... by... showed that...

(8)A study by... demonstrated that...

第 7 课　致谢(Acknowledgement)

致谢也称为鸣谢,位于正文后,参考文献前。对提供帮助或支持的个人或机构表示感谢,对同意在文中使用其知识产权的研究者表示感谢。

常用句型如下：

(1) We would like to thank... for helpful comments and discussions/technical support.

(2) The authors are grateful to... for providing...

(3) This work was supported by...

(4) Grateful acknowledgment is made to...

(5) The authors wish to express their sincere thanks to...

(6) Special gratitude is owed to...

(7) We would like to express our appreciation of/gratitude to...

(8) The authors thank... for giving financial support for this study.

(9) Additional support is provided by...

(10) Permission to quote from materials protected by copyright has been granted by...

(11) We thank... for the permission to quote from materials protected by copyright.

(12) Moreover, we are grateful to... for their valuable suggestions.

第 11 章

英文食品科技信息的获得

Acquirement of English Information of Food Science and Technology

第 1 课 主要的英文杂志

- ADVANCES IN FOOD NUTRITION
- AGRICULTURAL AND FOOD SCIENCE
- AGRICULTURE AND FOOD SECURITY
- AGROECOLOGY AND SUSTAINABLE FOOD SYSTEMS
- ANNUAL REVIEW OF FOOD SCIENCE AND TECHNOLOGY
- ANNUAL REVIEW OF NUTRITION
- ANTIOXIDANTS
- BIOSCIENCE OF MICROBIOTA FOOD AND HEALTH
- BRITISH FOOD JOURNAL
- CANADIAN INSTITUTE OF FOOD SCIENCE AND TECHNOLOGY JOURNAL
- CEREAL FOODS WORLD
- COMPREHENSIVE REVIEWS IN FOOD SCIENCE AND FOOD SAFETY
- CRITICAL REVIEWS IN FOOD SCIENCE AND NUTRITION
- CURRENT OPINION IN FOOD SCIENCE
- CZECH JOURNAL OF FOOD SCIENCES
- ECOLOGY OF FOOD AND NUTRITION
- EMIRATES JOURNAL OF FOOD AND AGRICULTURE
- EUROPEAN FOOD RESEARCH AND TECHNOLOGY
- FOOD ADDITIVES AND CONTAMINANTS
- FOOD ADDITIVES AND CONTAMINANTS PART B-SURVEILLANCE
- FOOD ANALYTICAL METHODS
- FOOD AND AGRICULTURAL IMMUNOLOGY
- FOOD AND BIOPRODUCTS PROCESSING
- FOOD AND CHEMICAL TOXICOLOGY
- FOOD AND COSMETICS TOXICOLOGY
- FOOD AND DRUG LAW JOURNAL
- FOOD AND ENERGY SECURITY
- FOOD AND ENVIRONMENTAL VIROLOGY
- FOOD AND FUNCTION
- FOOD AND NUTRITION BULLETIN
- FOOD AND NUTRITION RESEARCH
- FOOD BIOPHYSICS
- FOOD BIOSCIENCE
- FOOD BIOTECHNOLOGY
- FOODBORNE PATHOGENS AND DISEASE

- FOOD CHEMISTRY
- FOOD CONTROL
- FOOD CULTURE AND SOCIETY
- FOOD ENGINEERING
- FOOD ENGINEERING REVIEWS
- FOOD HYDROCOLLOIDS
- FOOD HYGIENE AND SAFETY SCIENCE
- FOOD MANUFACTURE
- FOOD MICROBIOLOGY
- FOOD PACKAGING AND SHELF LIFE
- FOOD POLICY
- FOOD QUALITY AND PREFERENCE
- FOOD RESEARCH INTERNATIONAL
- FOOD REVIEWS INTERNATIONAL
- FOODS
- FOOD SCIENCE AND BIOTECHNOLOGY
- FOOD SCIENCE AND HUMAN WELLNESS
- FOOD SCIENCE AND NUTRITION
- FOOD SCIENCE AND TECHNOLOGY
- FOOD SCIENCE AND TECHNOLOGY INTERNATIONAL
- FOOD SCIENCE AND TECHNOLOGY RESEARCH
- FOOD SCIENCE OF ANIMAL RESOURCES
- FOOD SECURITY
- FOOD STRUCTURE-NETHERLANDS
- FOOD TECHNOLOGY
- FOOD TECHNOLOGY AND BIOTECHNOLOGY
- INNOVATIVE FOOD SCIENCE AND EMERGING TECHNOLOGIES
- INTERNATIONAL FOOD AND AGRIBUSINESS MANAGEMENT REVIEW
- INTERNATIONAL FOOD RESEARCH JOURNAL
- INTERNATIONAL JOURNAL OF FOOD ENGINEERING
- INTERNATIONAL JOURNAL OF FOOD MICROBIOLOGY
- INTERNATIONAL JOURNAL OF FOOD PROPERTIES
- INTERNATIONAL JOURNAL OF FOOD SCIENCE AND TECHNOLOGY
- INTERNATIONAL JOURNAL OF FOOD SCIENCES AND NUTRITION
- IRISH JOURNAL OF AGRICULTURAL AND FOOD RESEARCH
- ITALIAN JOURNAL OF FOOD SCIENCE
- JOURNAL OF AGRICULTURAL AND FOOD CHEMISTRY
- JOURNAL OF FOOD AND DRUG ANALYSIS
- JOURNAL OF FOOD BIOCHEMISTRY

- JOURNAL OF FOOD COMPOSITION AND ANALYSIS
- JOURNAL OF FOOD ENGINEERING
- JOURNAL OF FOOD LIPIDS
- JOURNAL OF FOOD NUTRITION
- JOURNAL OF FOOD PROCESS ENGINEERING
- JOURNAL OF FOOD PROCESSING AND PRESERVATION
- JOURNAL OF FOOD PROTECTION
- JOURNAL OF FOOD QUALITY
- JOURNAL OF FOOD SAFETY
- JOURNAL OF FOOD SCIENCE
- JOURNAL OF FOOD SCIENCE AND TECHNOLOGY
- JOURNAL OF FOOD TECHNOLOGY
- JOURNAL OF MEDICINAL FOOD
- JOURNAL OF THE SCIENCE OF FOOD AND AGRICULTURE
- LWT-FOOD SCIENCE AND TECHNOLOGY
- MOLECULAR NUTRITION AND FOOD RESEARCH
- NUTRITION REVIEWS
- RENEWABLE AGRICULTURE AND FOOD SYSTEMS
- TRENDS IN FOOD SCIENCE AND TECHNOLOGY
- VETERINARY CLINICS OF NORTH AMERICA FOOD ANIMAL PRACTICE

第 2 课　主要的英文数据库

一、外文电子期刊数据库

- 外文国外食品安全专题数据库(OFSDE)
- 剑桥科学文摘网络数据库(CSA Expended Collection)
- 美国工程索引(EI)
- 美国化学文摘(CA)
- 美国生物医学信息检索系统 PubMed
- 新加坡电子期刊网 WordSciNet
- 斯坦福大学 Highwire Press
- 德国 Springerlink

二、专利数据库

- 美国专利数据库
- 加拿大专利数据库
- 欧洲专利数据库

第3课 专业英语学习的一些资源

- Science & Cooking：From Haute Cuisine to Soft Matter Science(physics)[科学和烹饪：从美味佳肴到软物质科学(物理学)]，美国哈佛大学
- Nutrition and Health：Micronutrients and Malnutrition(营养与健康：微量营养素与营养不良)，Nutrition and Health：Macronutrients and Overnutrition(营养与健康：大量营养素与营养过剩)，荷兰瓦格宁根大学
- Wine Tasting：Sensory Techniques for Wine Analysis(品酒：葡萄酒分析的感官技术)，美国加州大学戴维斯分校
- Introduction to Biology - The Secret of Life(生物学导论-生命的秘密)，美国麻省理工学院
- Child Nutrition and Cooking(儿童营养与烹饪)，美国斯坦福大学
- Medical Neuroscience(医学神经科学)，美国杜克大学
- Principles of Biochemistry(生物化学原理)，美国哈佛大学
- Meat(肉类食品)，美国佛罗里达大学
- Innovation：the Food Industry(创新：食品工业)，英国利兹大学
- Stanford Introduction to Food and Health(斯坦福大学食品与健康导论)，斯坦福大学
- Nutrition and Cancer(营养与癌症)，荷兰瓦格宁根大学

第 12 章

出国留学申请

Submit Application for Advanced Study Abroad

第1课 概 述

近年来，许多中国学生计划出国留学，进一步深造。申请出国留学能否获得成功，关键之一在于你能否向你希望进入的学校提供一份既实事求是又能充分展示自我的申请材料。

好的申请材料能使你从众多的申请者中脱颖而出，而平庸或较差的申请材料可能将你埋没。可以毫不夸张地说，在申请入学录取和经济资助时，申请材料在很大程度上和考试成绩一样重要。很多留学申请者没有充分认识到申请材料的重要性。有的人认为，只要我在TOEFL、GRE、GMAT、TSE或者LSAT之类的考试中取得了优异的成绩，就大功告成了。所以他们花大量的时间、金钱和精力去准备这样或那样的语言考试，而在准备材料时，往往草率行事，因此与出国深造的机会失之交臂。

一、申请材料的构成

不管申请什么专业，美国所有研究生院一般都要求以下材料。

1. TOFEL和GRE成绩

英语成绩的好坏对你的申请非常重要。只有具有相当程度的英语语言能力，你才有可能获得录取和资助。每所学校对TOEFL和GRE成绩都有自己的要求。大多数美国高校对申请食品科学专业考生的最低要求是：GRE为1 000分（词汇和数学两部分之和），TOEFL为80分，并且每一部分单项成绩不低于18分（新考试）。

2. 成绩单与学位证书

成绩单，即academic transcript。通常，研究生招生委员会非常看重你最近2年的主要课程的平均绩点。一般学校的研究生院要求平均绩点为3.0，但也有些学校会要求更高的绩点。招生委员会都会通过查看你的平均绩点和标准考试的成绩（如TOEFL和GRE）来评价你的申请。如果你的绩点很低，那你就有必要努力在GRE考试中取得高分。

如果你大学尚未毕业，可只提供现有的成绩单。所有成绩单都应当是中、英文对照的，最好2份同时盖上公章并封入信封，在封口上再盖上封签章。每所大学都会要求1～2份这样的成绩单。如果你读过研究生，自然也包括你在研究生学习期间的成绩单，这些成绩单的制作要求与本科成绩单相同。

3. 个人陈述

个人陈述，即personal statement或statement of interest。这是一份各专业学生都需要的自我陈述材料，一般要求500～1 000字。在这篇十分重要的文章中，你既要把对所学专业的精深看法写出来，还要把自己的追求和对前途的看法写出来。既要写得具体、实在，又要写得生动感人。

对于如何书写一份深思熟虑的个人陈述，中国学生在这方面并不擅长。个人陈述可以说明你的文化背景、出生地、专业发展经历、在生活或研究领域中的思考和信念、以后的打算等。一篇个人陈述不用包含以上所讲的全部内容，但你的个人陈述至少接近以下内容的一部分，比如讲个少见的故事，使人难忘，能表达一种精神。这是很重要的。它可以反映你的逻辑思维、写作能力、专业技能的水平。一定要用事实和例子来支撑你的陈述。如果你有引以为荣的东西，如你所上的名牌大学或自己的作品，那么一定要讲出来，表现出你可以在平时观察中总结出很多生活和学习的规律，这样你的陈述会很具有说服力。

如何写这份材料，也是本章的重要内容。

4. 推荐信

推荐信，即 reference 或 letter of recommendation。美国的学校一般要求每个申请者提供 3 封推荐信。大部分学校都有固定的推荐信表格，只要按照表格填写就行。多数情况下，推荐信也可以由推荐人另行用信纸写成，单独或随表格一起寄给国外学校。每份推荐信都必须由推荐人亲自签名，不能复印。推荐人最好由了解你的大学老师、单位领导及专业知名人士担任。如果有外国教授比较了解、欣赏你，请他们写推荐信，效果往往不错。

5. 申请说明函

申请说明函，即 cover letter。在提交申请材料时，你需要写一份简短的说明函对你的申请材料和你喜欢的学术领域进行说明，并让招生负责人知道你的一些重要信息。与这些人交流时一定要礼貌，内容明确。说明函一定要围绕你想进入的研究生课程进行，不要用千篇一律的说明函。

6. 补充材料

在递交申请材料时，还可以根据自己的情况提供如下的补充材料，如资金证明表、简历、论文摘要或项目设计介绍，以及所获得的各种荣誉与奖励证书等。

二、申请流程

1. Submit application 提交申请

——Most universities offer online applications 大多数大学提供网上申请

2. Graduate school evaluation 研究生院评估

——TOEFL and GRE scores TOEFL 和 GRE 成绩(是否达到要求)

——GPA 平均绩点(是否达到要求)

——Document completion 申请文件(是否齐全)

——Application fee 申请费用(是否支付)

3. Department admission committee evaluation 系招生委员会评估

——TOEFL and GRE scores TOEFL 和 GRE 分数

——GPA 平均绩点

——Personal statement 个人简述

——Reference letters 推荐信

4. Admission decision 招生决定

三、研究生奖学金

1. Graduate teaching assistantship (GTA) 教学助教奖学金

——Limited availability 名额有限

——Responsibility in teaching assistant in addition to the thesis research 职责包括助教及助研

——Usually 9 months appointment (without summer salary)一般是 9 个月委任期(没有

暑假工资）

2. Graduate research assistantship（GRA）研究助教奖学金

——Based on research projects of faculties 基于研究项目

——Research plan is usually designed as thesis research 研究通常设计成研究课题

——Most likely it will be 12 months appointment 大概为12个月的委任期

——Whether renewable or not depending on research grant 是否续聘取决于研究经费

3. Graduate assistantships vary among different universities 各学校间的研究生助教奖学金有差异

——Monthly salaries provided to support living expense 提供月薪供生活所用

——Tuition may be included or not 可能包括或不包括学费

——Health insurance is your responsibility and you have to pay for yourself 健康保险是你的责任，你必须为自己支付费用

第2课　个人陈述

个人陈述，即 personal statement，是所有出国文件中最重要的部分。它是申请者最主要的自我包装，把自己推销给评审者，从而达到申请被录取和获得经济资助的目的。中国申请者往往没有面谈的机会，因而这份文件愈发显得重要。我们要建议所有的申请者必须抓住机会，充分而又恰如其分地表现你的人格魅力。你的重任是让这一文件反映出你的个性和才智。写自述即使对以英语为母语的北美学生也是颇费心力的事情，对中国申请者来说，用英语写作本来就是困难的事，更何况自述是为了表现和包装自己，这与我们的文化习俗格格不入，写起来往往倍感费力。

一、个人陈述写作技巧

1. 个人陈述选择主题

(1)Make a list of all your experiences and interests first. 先列出你所有经历和兴趣。

(2)Find an overlapping theme or connection between different items on the list. 在列表中寻找重叠主题或者不同主题的衔接点。

(3)Your underlying theme should be why you should be accepted into the program to which you are applying for. 你的重点应该写你为什么应该被你申请的研究生院所接受。

(4)Your job is to sell yourself and distinguish yourself from other applicants through examples. 你的工作是推销你自己，并且通过实例使你有别于其他的申请者。

2. 使用第一人称

(1)Try to speak in the first person on your personal statement. 在你的个人陈述中尽量采用第一人称叙述。

(2)Your goal is to make your essay sound personal and active. 你的目标是确保你的文章听起来既有个性又生动。

(3)Avoid overusing “I” and, instead, alter between “I” and “me” or “my”. 避免滥用

"I"，而应该在"I"和"me"或者"my"间转换。

3. 关于个人陈述中研究兴趣的讨论

(1)First and foremost，be specific about your research interest in your personal statement. 首先最重要的是，在个人陈述中你的研究兴趣要具体。

(2)You may describe research interest in a few areas，such as food microbiology，safety，and biotechnology. 你可以在某些领域描述你的研究兴趣，如食品微生物学、食品安全和食品生物技术。

(3)Your aim is to show your readers that you have knowledge in your proposed field of study. 你的目的是向读者展示你在自己的研究领域已掌握的知识。

(4)Do not forget to describe your academic goals. 别忘记描述一下你的学术目标。

(5)A clearly stated research interest allows admission committee to pass your application to faculty members whose primary research matches your interest. 清晰的研究兴趣能使招生委员会将你的申请材料转给与你研究兴趣相匹配的教师。

(6)A general interest in food science and technology should be avoided. 应该避免只说对食品科学与工程学科感兴趣(不具体)。

4. 没有特殊经历或技能时的表述

(1)Make a list of all your qualities and think of how you utilized them in the past. 罗列出你所有才能，并回忆你过去曾如何发挥这些才能。

(2)Discuss the ones that will make you stand out and have connection to your field of interest. 讨论那些将使你突出并与专业兴趣有关的才能。

(3)If you do not have many experiences within your field，then try to make your other experiences relate to your interests. 如果你在该领域没有多少经历，那你就尝试让其他的经历同你的研究兴趣有联系。

5. 选择指导教师

(1)Yes. It makes it easier for the admission committee to determine if your interests match with the faculty members you are interested in working with. 是的。这使招生委员会更容易判断你的研究兴趣是否与你想共事的教师相吻合。

(2)Try to mention more than one professor you wish to work with because sometimes there is a possibility that the professor you are interested in working with is not accepting new students for that year. 尽量指出 2 个以上你希望一起共事的教授，因为可能你感兴趣的教授那年不招收新学生。

(3)By mentioning only one professor，you are limiting yourself，which can decrease your chances of being accepted. 如果你仅提及 1 位教授，会使自己受到限制，那将减少你被录取的机会。

6. 个人不足之处的分析

(1)If you think it may be helpful，then you should discuss low grades，low GRE scores，etc.，providing an explanation. 如果你认为可能有帮助，那你应该就你的低成绩、低 GRE 分数等进行讨论说明。

(2)Do not blame others, or try to explain away three years of poor performance. 不要指责其他人,或者试图对过去3年的不良表现进行辩解。

(3) Provide explanations that are reasonably excusable and comprehensive to the academic committee, such as an unexpected accident in the family. 向学术委员会提供合理和全面的解释,诸如家庭发生的意外遭遇。

7. 个人陈述的篇幅

(1)Generally, there is no limit of words in the personal statement, but sometimes it depends on schools and programs. 一般来讲没有限制,但取决于学校和研究项目。

(2)Ideally, a personal statement is between 500 to 1 000 words long. 一篇理想的个人陈述在500～1 000字。

(3)Remember to cover all important factors. 记住要涵盖所有重要的因素。

(4)A lengthy statement does not always catch the eyes of the admission committee. 一篇冗长的陈述并不总能吸引招生委员会的眼球。

二、个人陈述的结构和组成

(1)There are different ways you can structure your essay but the most common format includes: (1) introduction, (2) main body, (3) conclusion. 你可以采用多种形式来组织你的文章,但是最通用的格式包括:前言、主体和结论。

(2)Your essay should include enough details. Be personal and specific and show what makes you unique and different from other applicants. 你的文章应该包括足够多的细节,要有个性,要显示出你比其他申请者独特的一面。

1. 前言

(1)An introduction is the most important part of the essay, especially the first sentence. 前言是整篇文章最重要的部分,特别是第一句话。

(2)First sentences may explain your desire to study the subject of interest or discuss the motivation that influenced your desire to study the subject of interest. 开首语可以表述你对所感兴趣的项目的学习愿望,或者讨论影响你学习兴趣的活动。

(3)The sentences following the first sentence should provide a brief explanation that supports the claim stated in the first sentence. 紧接着第一句话应该提供简单的解释来支撑第一句话。

2. 主体

(1)The body should include several paragraphs (usually about 3) that provide detailed evidence to support the statement made in the introductory. 主体部分应该包括几段(一般为3段),提供一些具体证据来支撑介绍部分所述的观点。

(2)Each paragraph should have a topic statement, which ends each paragraph with a meaningful sentence that provides a transition to the next paragraph. 每个段落应该有一个话题,其结束需要有个能够启下的语句。

(3) Experiences, accomplishments, future goals and any evidence that can support your application should be included in the body. 经历、成果、目标以及任何能支撑你申请的

证据都应该包括在主体里面。

(4)A short summary of your educational background can be discussed in the 1st paragraph. 对于你教育背景的简单陈述可以在第一段里进行。

(5)Personal experiences and the reasons for wanting to attend the school can be discussed in the 2nd paragraph. 个人经历和入学原因可以在第二段里讨论。

(6)The last paragraph should explain why you should be accepted. 最后一段应该阐述你为何应该被录取。

3. 结论

(1)The conclusion is the last paragraph of the personal statement. 结论是个人陈述的最后一段。

(2)State why you are interested in studying the subject of interest. 说明你为何对所研究的课程感兴趣。

(3)State the key points mentioned in the body, such as your experiences or accomplishments that explain your interest in the subject. State it conclusively and briefly. 阐述主体中提及的重要部分,如你的经历或者成果来解释你对该研究主题的兴趣。要用简单明了的方式来阐述。

(4)End on a positive note with one or two attention-grabbing sentences. 以 1～2 个吸引眼球的话语作为结束语。

三、个人陈述注意事项

(1)Prepare an outline and create a draft. 准备提纲并写草稿。

(2)Make sure your essay has a theme or a thesis. 确保你的文章有主题。

(3)Provide evidence to support your claims. 提供证据来支撑你的观点。

(4)Make your introduction unique. 使你的前言具有独特性。

(5)Write clearly and make sure it is easy to read. 文章要清晰,通俗易懂。

(6)Be honest, confident, and be yourself. 诚实,自信,有个性。

(7)Make sure your essay is well-organized and concise. 确认你的文章组织严谨,简明扼要。

(8)Discuss your future goals. 讨论你将来的目标打算。

(9)Mention any hobbies, past jobs, community service, or research experience. 提及兴趣爱好,过去的工作,社会服务或者研究经验。

(10)Use examples to demonstrate your abilities. 例证你的能力。

(11)Proof read. (拼写)校对。

四、个人陈述应避免事项

(1)Have any grammar or spelling errors. 语法及拼写错误。

(2)Use jargon or slang. 用方言土语或俚语。

(3)Be boring or repetitive. 枯燥重复。

(4)Generalize. 泛泛而谈。

(5)Be comical (a little humor is okay but remember it can be misconstrued). 滑稽(一

点点幽默是可以的，但是记住它可能被误解)。

(6)Give excuses for a low GPA. 为低 GPA 找借口。

(7)Make lists of accomplishments, awards, skills, or personal qualities (Show, don't tell). 罗列出成果、奖励、技能或者个人品质。

(8)Write an autobiography or summarize your resume. 写自传或概述你的简历。

(9)Include information already cited on the application. 包括已经在申请中提及过的信息。

(10)Forget to proofread. 忘记校对。

五、个人陈述范文

1. 申请食品安全方向硕士研究生的个人陈述

Personal Statement
Proposed Major: Food Science
Proposed Specialization: Food Safety

This is a typical day of a Chinese man: "He gets up in the morning and brushes the teeth with cancerous toothpaste; feeds his baby with milk powder containing excessive amounts of iodine; pours himself a glass of milk he bought yesterday not knowing it was wrongly dated; eats a bun that is too white to be true and munches on pickles make at a stinking roadside ditch. In the afternoon, he joins his peers at the nearest KFC outlet for a lunch of fried chicken containing the chemical Sudan Ⅰ, a dye that causes cancer in mice and rabbits. Before he leaves the office at the end of the day, he calls home where his wife is preparing a dinner of chemically contaminated rice, pesticide-infested vegetables cooked with recycled oil". The shocking story, although not completely true, is the one prevailing on the Internet, which illustrates that we are facing many food problems.

Therefore, I am eager to pursue my graduate study with a major in food science, particularly in the area of food safety. I feel that, by studying and researching at an advanced level, I can both expand on my existing knowledge and apply them to solve the food safety problems in my country. Furthermore, to a large developing country like China, the demand for talents in various fields is tremendous. Thus, I wrote this statement to apply to the graduate program in the Department of Food Science of Oregon State University as your well-equipped laboratories and distinguished faculty can provide good environments for my research. For example, Dr. Yi-Cheng Su of your department, he is a visiting professor in our university and used to give us a lecture on seafood safety. I am very interested in his research area and hope to join his research group.

Some may ask me why I turn to the major of food science, whereas my undergraduate major is biotechnology, which aims to apply biotechnology to the pharmacy industry? To begin with, foods are the necessity for all people while not everyone needs to take medicine three times a day. So I think I can help more people with their food problems if I

am qualified with certain certificates. Secondly, I have finished compulsory courses such as organic chemistry, microbiology, biochemistry, molecular biology which provide the prerequisites for future study of food science. Finally, my educational background is superior to other applicators as I have taken courses like physiology, pharmacology, basic immunology, utilization of marine resources, which can connect the fields of pharmacy and food science.

To fulfill my dream of engaging in the works related to food science, I take every opportunity to be familiar with this area. At first, I have taken the course of Introduction to Food Science and Technology and passed the examination with flying colors. I also took the training course of HACCP in my leisure time and then got the certificate. Besides, I took an internship in the Food Quality Supervision & Inspection at Shanghai Station in the P.R.C. Light Industry Ministry in two summers. I was instructed to operate some devices (such as UV-visible spectrophotometer, GC, fluorescence spectrometer etc.) to detect constituents in foods. For instance, I used the Griess Reagent to detect the quantity of nitrite in meats and vegetables; I detected the quantity of vitamin C and vitamin B_1 in foods with the aid of devices; I also observed the chromaticity and turbidity of drinking water according to the relative documents. During those periods, I have not only learned many detection methods but also realized our country's food safety challenge, which reinforces my determination to further my education in the major of food science. Currently, I am participating in a research project "Detection and analysis of paralytic shellfish poison (PSP) and diarrhetic shellfish poison (DSP) in red tide algae by HPLC".

If I have the opportunity to be trained here, it is a good way to know about food issues, not only know "how" but also know "why". In the laboratories, I will cherish every moment to take advantage of advanced facilities to research according to theoretical knowledge. Meanwhile, studying aboard is a good chance to meet new people, understand the new culture, and experience new lifestyles. After my graduation, I will apply all gained knowledge and research abilities to my future possible careers. I would like to be a college teacher especially in my alma mater or a specialist in the inspection organization or industry. Anyway, I am willing to do something to improve our country's food safety.

2. 申请葡萄酒制造方向博士研究生的个人陈述

Personal Statement

Proposed Major: Food Science

Proposed Specialization: Wine and Winegrapes

This letter is submitted to the FST graduate committee to convey my intentions to pursue a Ph. D. in Food Science and Technology at Oregon State University.

FST Graduate Committee,

After spending many years between wineries and academic research, I determined the world of scientific and academic research was a fit for me. I came to OSU three years ago with the intent of obtaining a Ph. D. through the study of wine and winegrapes. During those years I have developed the research skills and discipline required to be successful. I recently proved this with the defense of my Masters' thesis (September 13th, 2007) highlighting independent research I have completed to date. My thesis consists of two manuscripts that have been submitted to high-impact journals (*Critical Reviews in Food Science and Nutrition and Analytica Chimica Acta*) and will be reviewed within the year to come. I was also invited to give an oral presentation at an international symposium for analytical chemists studying grapes and wine (*In Vino Analytica*), which was very well received. My research has been funded following the submission of a grant proposal, research update, and renewal that I co-wrote with my advisor, Dr. James Kennedy.

During this time I have excelled in all coursework at OSU with a current GPA of 3.75. I have also been a TA for courses in wine production and analysis as well as sensory evaluation. For my part in the wine sensory class, I received an outstanding TA award, for which I am grateful. I have been a leader of the OSU Vitis Wine Club as well as the Food and Fermentation Club and acted as the graduate student representative at Faculty and Graduate Committee meetings. I believe that all of these experiences have provided me with a realistic glimpse of academic life as well as life as a research scientist in a field I am incredibly passionate about. I have experienced various roles of leadership, team building, personal and group development, as well as one's role in the scientific community. I have been exposed to the process of writing successful research grants, conducting rigorous scientific research and reporting findings in person and in peer-reviewed journals.

Most importantly to me, I have been approaching my current research as a multi-year study worthy of a Ph.D. My research has a primary focus on providing enough information to make a significant contribution to the fundamental science I study. My intention was not to take the necessary classes and fulfill the duties required to obtain a degree, rather to conduct quality research and make a reputation for myself in my community. My goal of obtaining a Ph.D. is such that in the future I will have the option of carrying my work, guidance and teaching on in a university setting where I may be able to help others along the way. I hope you realize the intention of my studies here at OSU and see fit to allow me to continue my research, for which I have secured funding. I look forward to completing my research and continuing to elevate the reputation of OSU in the scientific research community.

Sincerely,
Seth Cohen

3. 申请功能性食品方向硕博连读研究生的个人陈述

Personal Statement

Proposed Major: Food Science

Proposed Specialization: Functional Foods

Although there is no official definition of functional foods, it is generally considered that they are a group of foods that provide physiological benefits beyond those traditionally expected from food. There is no denying that, with the development of society, people become increasingly interested in this kind of food. In my humble opinion, functional foods are more than the foods to support our life, they are part of a healthy life style. That is why I am so eager to pursue my graduate study in this area, not only to improve the health condition myself, but also to help more people cultivate better dietary habits. As for me, Cornell University is absolutely my the first choice. One reason is that, it has a high reputation in food science and also being the first university to establish the area of functional foods. Another reason is your excellent faculty member-Rui Hai Liu, because his research interests are suitable for my academic background and interests.

My undergraduate major on Biotechnology (Marine Pharmaceutical Technology) this has provided the prerequisite for my future education in food science, particularly in the area of functional foods, in that we are going to apply biotechnology to the pharmacy industry. The major courses I took include Anatomy and Physiology, Pharmacology, Chemistry of Natural Drugs, Pharmacy Analysis, Basic Immunology, Utilization of Marine Resources etc. I learned a lot related to these courses in the laboratory. For instance, I learned to identify the constituents (such as rutin and guereetin) in a Chinese herb *Flos Sophorae Immaturus* with TLC (thin layer chromatography) method. I learned to do the animal test to observe the influence of drugs on them. For example, I observed different responses of mice to drugs administrated by different routes include intraperitoneal injection, muscle injection, and tail veil injection. I also analyzed the cardiogram of mice both taking medicine and those not taking. I learned to identify aspirin by chemistry reaction, test the general impurity and specific impurity and then determine the content of aspirin by titration method. By winning scholarships every academic year, I am confident of bridging the area of pharmacy and food science in my graduate study.

In order to know about the foods more deeply, I also took the courses of Nutrition. In the courses, I was told some functions of certain kinds of foods. For example, foods containing beta-carotene (carrots, pumpkin etc.) can neutralize free radicals; foods containing lutein and zeaxanthin may contribute to the maintenance of healthy vision; insoluble fiber may contribute to maintain a healthy digestive tract and reduce the risk of some types of cancer. However, what the courses told me is just the beginning, which stimulates my interest in exploring the beneficial compounds from nature.

Currently, I am a research assistant at East China Sea Fisheries Research Institute, participating in the research project "Isolation and purification of natural products from marine dinoflagellates". We are now on the way using the rotary evaporator to isolate the compounds from *Alexandtrium minutum* (Amtk-4, Amtk-9) and *Prorocentrum lima* (PL01, PL03) and waiting to identify the characters of the natural products. Research processes sometimes are trivia and boring, but thinking that we are trying to find new natural products, which may be the useful ones in food or medical aspects, all efforts deserved.

If I am admitted, I will strive for achieving the following goals: (1) To grasp the theoretical knowledge solidly, especially the basic knowledge on the text books; (2) To advance my understanding of functional foods, which means I should keep up with latest information in relative areas; (3) To improve the research abilities, which means I can take advantage of different kinds of devices to operate my desired experiments.

After graduation, I would like to be a specialist in the academic institute, become solidly grounded in the methods of separating, analyzing, and identifying chemical compounds in Chinese herbal medicine and other plants. Our country owns many precious and rare herbs and plants, so I want to be a member of defining their characters and functions.

第3课 推 荐 信

一、概述

A well written letter of recommendation provides the admission committee with information that is not found elsewhere in the application. A detailed discussion, from a faculty member, of the personal qualities, accomplishments, and experiences will make you unique and perfect for the programs to which you have applied for.

一封优秀的推荐信将向招生委员会提供你申请材料中看不出来的信息。大学教授对你的个人品质、成果及经历等做详尽评价,使你与众不同,成为申请院校研究生项目中最完美的人选。

二、寻找推荐者

(1) Know you well 对你很了解

(2) Know you long enough to write with authority 认识你足够长的时间,从而使推荐信具有权威性

(3) Know your work and evaluate your work positively 了解并能对你的工作(表现)给予积极评价

(4) Have a high opinion of you 对你有高度评价

（5）Know your educational and career goals　知道你教育及职业生涯的目标

（6）Be well known　著名的

（7）Be able to write a good letter　能写出一封好的推荐信

记住，没有人能满足以上所有的标准。理想情况下，推荐信应涵盖你的学术水平、学习技能、研究能力和工作经验以及见习经历（如实习以及相关工作经验）。

三、需要向推荐者提供的信息文件

（1）Courses you've taken with 你参加过的课程

（2）Transcript and resume 成绩单和简历

（3）Admissions essays 入学申请书

（4）Research and work experiences 研究及工作经历

（5）Internship and other practical experiences 实习或其他实际经历

（6）Honor of societies to which you belong 你所在的社会团体的荣誉

（7）Awards you have won 你所获的奖励

（8）Professional goals 职业目标

（9）The due date for the application 申请截止日期

四、推荐信范文

1. 申请去国外作交流学生的推荐信

December 1, 2020

To whom it may concern,

As Head of the College of Food Science and Technology of Shanghai Fisheries University, I am pleased to recommend Mr. ________ for acceptance into your program as an exchange student.

Mr. ________ is one of my favorite students, has attended my class of Marine Bioresources Utilization gently. He always takes a front seat when he attends the class. He shows great interest in this course. He is very active with a quick response in the class. All of those give me a deep impression.

Mr. ________ is a talented young man who has done an excellent job in his study. In 2018, He was admitted to our department because of his high scores in the National College Entrance Examination. During the past two years as an undergraduate student, he has been granted 99 credits with a 3.5 GPA and passed the test for C Language of Computer Science (Band two) and College English Test (Band 6, highest band for non-English major students). Due to his excellent performance in his study, he has been awarded Student Scholarship for several times (First Prize, twice; Second Prize, once). In addition, he is also active in many social activities, Charity of Children volunteer, Science and Technology Innovation Club member of Shanghai Fisheries University.

Mr. ________ is also responsible, honest, and warm-hearted. His pleasing personality makes him very popular in both his class and our department. I believe that his maturity and intellect will enable him to adapt himself quickly and easily to any environment. Therefore, I strongly recommend him to you for consideration and accept him as an exchange student in your lab.

Your kind consideration will be greatly appreciated. I would like to provide more information upon your request. You can reach me by the home phone and mail address or via the Internet. My email address is xxxxx@shfu.edu.cn

Sincerely,
×××
Address
Telephone number
Fax number
E-mail

2. 申请美国高校硕士研究生的推荐信

November 6, 2006

To whom it may concern,

It is a real pleasure to recommend Miss ________ to you for admission to the graduate program at your university.

Miss ________ was one of my students who attended my classes and did a research project for her bachelor thesis under my guidance. I have known her since she became an undergraduate student at the College of Food Science and Technology, Shanghai Fisheries University. I was deeply impressed by her good command of English especially oral English. She used to chair several university parties in English.

As an undergraduate student in my class, Miss ________ academic performance did not rank the top in the first year. At the beginning of her undergraduate study, her knowledge was as limited as most of the other students. But she quickly grasped the new knowledge through hard-working when she realized the importance of courses. She often studied even during weekends and holidays and greatly enjoyed new findings, new ideas, raising questions, and solving problems.

During the lab work for her thesis, I noticed her strong desire to learn more experimental techniques on pharmacology and sent her to conduct a collaboration project with Professor Fan, director of Laboratory of Medical Analysis, School of Pharmacy, the Second Military Medical University. Her bachelor thesis was titled "*Assessment of intestinal absorption of vitexin-2-O-rhamnoside in hawthorn leaves flavonoids in rat using in situ and in vitro absorption models*". She worked more independently and did a better job compared with most of her classmates. In addition, she performed excellently as well during her thesis defense.

From all my knowledge, I judge that Miss ________ is an active and promising student and very cooperative with other people. Therefore, I willingly recommend her to you for consideration and accept her as a candidate for your graduate students.

Please feel free to contact me if you need more information.

Yours sincerely,

×××

3. 申请英国高校博士研究生的推荐信

May 15, 2006

To whom it may concern,

Dear Sir,

Miss ________ requested a letter of recommendation in support of her application for the doctoral program at Kings College. As an emeritus professor of the department of Food Science of Shanghai University, I am pleased to comply with her request.

Miss ________ was one of the students who attended my lecture in Food Nutrition during her undergraduate studies at Shanghai Fisheries University. I was deeply impressed by her hard work and academic performance. At the beginning of the lecture, her background on food nutrition was very limited, just as the same as other students. But she quickly grasped the new knowledge through hard-working. She often studied even during weekends and holidays and greatly enjoyed new findings, new ideas, raising questions and solving problems. In the final examination, she got Grade A and ranked among the top 3 of the class.

In addition, Miss ________ was also active in other activities. I was once invited to be a judge for a competition for English speaking in 2000. As a freshman among upper-grade competitors, she showed wit and self-possession. Her clear voice, accompanied by a vivid expression, made her exceptional among her peers. She won the championship. I heard Miss Name took part in speaking competitions in the following semesters and won a prize every time.

On a personal level, Miss ________ is courteous, pleasant, and helpful. She made many friends among her classmates. During her study of master course at the University of Aberdeen (UK), Miss ________ contacted me occasionally through e-mails and phone calls and consulted me about her future plan for further study in the doctoral program. I am pleased when she told me she would like to pursue advanced studies in your program. Judging by what I know of her, she is fully qualified to pursue a Ph. D. in your program. Therefore I recommend her with strong enthusiasm.

I would greatly appreciate it if you could give her application for candidacy to your Ph. D. program with favorable consideration. Please feel free to let me know if you need additional information about her application.

Truly,

×××

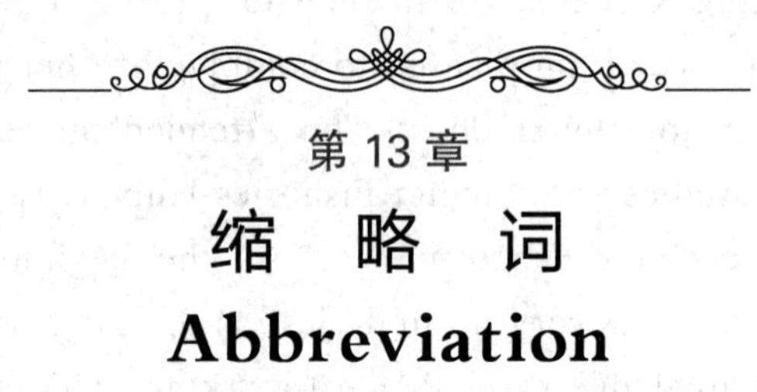

第 13 章

缩 略 词

Abbreviation

第 1 课　食品加工与贮藏类

ADL (acceptable defect level)容许缺陷标准
A.E. (air entraining)充气,加气
AE (absolute error)绝对误差
Ae (air escape)漏气,放气
AFD (accelerated freeze-drying)加速冻干
AGV(automated guided vehicle)自动导航车
Ahr (acceptable hazard rate)容许公害率
ALICE(automatic leaf inspection and conveyer)烟叶自动检查和输送器
alky. (alkalinity)碱度
approx. (approximately)大约,近似
AQL(acceptable quality level)合格质量极限
ATCT(alloy-tin couple test)合金-锡电偶试验
atm. (atmosphere)大气压
*A*w(water activity)水分活度
B&C(bakery and cookery)(面包)烤制和烹调
BFC(baby food cap)婴儿食品罐盖
BHPP(batch high pressure processing)间歇式高压加工
BOD(biological oxidation demand)生物需氧量
C.I. (color index)比色指数
CA ①(compressed air)压缩空气;②(controlled atmosphere)快速降氧法
CAS(controlled-atmosphere storage)气调储藏
CCP(critical control point)临界控制点
CE(calibration error)校准错误
CFC(chlorofluorocarbon)氯氟烃
CHPP(continuous high pressure processing)连续式高压加工
CHU/P(caloric heat unit/pound)摄氏热单位
CI ①(clearing index)净化指数;②(concentration index)浓度指数
CIP(cleaning-in-place)原位清洗
CIP system(clean-in-place system)原位清洗系统
COD(chemical oxidation demand)化学需氧量
CORBA (common object resource based architecture)共同目标资源软件
COV(coefficient of variation)便宜系数
CPF(capsule packed freezing)冰壳冻结法
CPM(can per minute)每分钟罐数
CRS(cold-rolled steel)冷轧钢

D.S. and S.P. (dry salted and smoked products)干腌熏制品，腊制品

DCS(distributed control system)集散式控制系统

DDE(dynamic data exchange)动态数据交换

DE(dextrose equivalent)葡萄糖当量

DF(diafiltration)渗滤

DH(degree of hydrolysis)水解度

DI can(drawn and ironed can)冲拔压薄罐

DI (dairy industries)牛奶工业

DRD(drawn-redrawn)多级冲拔罐(直径接近高度)

DS ①(dilute strength)稀释程度；②(dry substance)干物质(质量分数)

DT. (dry top)汤汁不足致使固形物露出液面(罐头检验术语)

DTO(deep twist-off cap)深型四旋盖，深型爪旋盖

DV(dilute volume)稀释容积

DW(distilled water)蒸馏水

EDI(electronic data interchange)电子数据交换

ETP(electrolytic tinplate)电镀锡薄板

EU (entropy unit)熵单位

EWL(egg-white lysozyme)卵清溶菌酶蒸发单位(在 100 ℃及每平方英尺 1.17 磅压力下蒸发 1 磅水所需热量)

FFA (fresh freezer accumulation)冷库近 15 d 内进肉量

FID(flame ionization detector)氢焰离子化检测器

FP (freezing point)冰点，凝固点

FPM (feet per minute)英尺/分钟

GMP(good manufacture practice)良好操作规范

GWP(global warming potential)全球变暖潜能值

HACCP(hazard analysis critical control point)危害分析及关键控制点

HCF method(heat-cool-filling method)杀菌-冷却-装罐法

HDP(hot-dipped plate)热浸镀锡薄板

HDTP(high-density tinplate)热浸镀锡薄板

HHP(high hydrostatic pressure)高静水压技术

HIT can(high tin fillet can)高锡带罐(罐身接缝内侧焊有高纯锡条以抗罐壁氧化)

HPF(homonizing process freezing)均温冻法

HPP(high pressure process)高压处理杀菌

HQL(high quality life)高品质寿命

HTST(high temperature short time)高温瞬时

I.H.P. (indicated horse power)指示马力

I.Q.F. (individually quick frozen)个体速冻

I.S.F. can(inside solder fillet can)罐内焊锡技术

IBC(intermediate bulk container)中型集装箱

ID-CA(two-dimensional dynamic controlled atmosphere)双相变动贮藏
IQB(individual quick blanching)单块快速烫漂
IQF(individual quick freezing)单块快速冷冻
IS machine(individual section machine)制玻璃瓶用,行列机
ISV(iron solution value)(镀锡板的)铁溶出值
IT(initial temperature)初温
JIT(just-in-time)看板作业
L.N.F. (liquid nitrogen frozen)液氮冷冻
MA(modified atmosphere)自然降氧法
MAS(modified-atmosphere storage)气调贮藏
MRP(material resource planning)物质资源计划
MTS(mano-therno-sonication)压-热-声处理
N.F.S. (not fat solid)非脂肪固形物
N.T.P. (number of theoretical plate)理论板数
N/N(not to noted)无标记
NFE(nitrogen-free extract)无氮抽提物
NHV(net heating value)净热值
Nt/Wt(net weight)净重
ODBC(open database connectivity)开放式数据库连接
ODP(ozone depletion potential)大气臭氧层消耗的潜能值
OHTC(overall heat transfer coefficient)总传热系数
OLE(object linking and embedding)目标程序连接和嵌入
p.u.p. (pick-up-pump)真空泵
PCS(process control systems)过程控制系统
PEP(pulsed electric field)脉冲电场技术
PLC(programmable logic controllers)可编程逻辑控制器
PLV(pickle lag value)(镀锡板的)酸浸时滞值
PPP(product-processing-packaging)产品生产过程包装
PRAC(vacuum pre-refrigerated atmospheric control system)真空预冷气调保鲜系统
PRHS(pre-refrigerated hypobaric storage)真空预冷减压气调贮藏
PSL(practical storage life)实用冻藏期
QFF(quick frozen foods)速冻食品
Q-F(quick-freezing)速冻法
R.I. (refractive index)折光率,折光指数
RDA(recommended daily allowance)推荐日摄入量
REH(equilibrium relative humidity)空气相对湿度
REM(roentgen equivalent man)人体伦琴当量,生物伦琴当量
REP (roentgen equivalent physical)物理伦琴当量
REPFED(ready-to-eat-products-for-extended-durability)可长期贮藏的即食产品

r/min(revolutions per minute)转每分
RH(relation humidity)相对湿度
RO(reverse osmosis)反渗透
RP(retort pouch)软罐头
r.p.m. (revolutions per minute)每分钟转数
RQ(respiratory quotient)呼吸商
RV(residual volume)残余容积
SA(spectrum analyzer)光谱分析器
SC(super quick chilling)超级快速冷却
SCADA(supervisory control and data acquisition)监督控制和数据获得
T cap ①(pano-T cap)滚轧螺纹盖;②(press-on twist-off cap)热成形螺纹盖
TDS(total dissolved solid)溶解总固体
TDT(thermal death time)热力致死时间
TEWI(total equivalent warming impact)总等效温室效应
TFS-CT(tin free steel chromeplate treatment)镀铬薄板
THA(thermal hysteretic activity)热滞活性
TQM(total quality management)全程质量控制
TRT(thermal reduction time)热力指数递减时间
TTT(time-temperature tolerance)时间-温度耐受性
U.H.T.S. (ultra high temperature sterilization)超高温杀菌
UF(ultrafiltration)超滤法
UHS(ultra-high speed)超高速
UHV(ultra-high vacuum)超高真空
UL(upper limit)上限
ULO(ultra low oxygen)超低氧贮藏
UTH(ultra high temperature)超高温
V.G(viscosity gradient)黏度梯度
VCID(vacuum cooling infrared-rays dry system)真空冷却红外线脱水保鲜技术
WAI(water absorption index)吸水系数
WOF(warm-over flavor)陈腐的风味
WSC(water solubility characteristic)水溶特性
WWT(waste water treatment)废水处理

第2课　食品微生物学及生物工程类

ACP(acyl carrier protein)酰基载体蛋白
ADP(adenosine-5′-diphosphate)腺苷-5′-二磷酸
AMP(adenosine-5′-mono phosphate)腺苷-5′-磷酸

APS(ammonium persulfate)过硫酸铵
A-site(amino-acyl site)氨酰基(A)位点(核糖体)
ATP(adenosine-5′-triphosphate)腺苷-5′-三磷酸
ATPase(ATP synthase)腺苷三磷酸酶
B.O.D. (biological optical density)细菌光密度
BMI(bovine milk isozyme)牛乳同工酶
BOD_5(five day BOD)5 日生物需氧量
bp(base pair)碱基对
CAMP(cyclic AMP)环腺苷酸
CAP(catabolite activator protein)分解代谢物激活蛋白
CAT(chloramphenicol acetyl transferase)氯霉素乙酰转移酶
CDA(chitin deacetylase)甲壳素脱乙酰酶
CFU(colony-forming unit)菌落形成单位
CMP(cytidine-5′-monophosphate)胞苷-5′-磷酸
CMV (cytomegalovirus)巨细胞病毒
CNS(central nervous system)中枢神经系统
CoA(coenzyme A)辅酶 A
CPE(cytopathic effect)细胞致病作用
C-phase(chromosome replication phase)染色体复制期
CRP(cAMP receptor protein)cAMP 受体蛋白
CrP(creatine phosphoric acid)肌磷酸,肌酸磷酸
CTL(cytotoxic T lymphocyte)细胞毒性 T 淋巴细胞
CTP(cytidine-5′-triphosphate)胞苷-5′-三磷酸
DAP(meso-diaminopimelic acid)内消旋二氨基庚二酸
DCW(dry cell weight)细胞干重
ddH_2O(distilled deionized water)蒸馏去离子水
DHA(dihydroxyacetone)二羟丙酮
DMSO(dimethylsulfoxide)二甲基亚砜
DNA(deoxynucleic acid)脱氧核糖核酸
DNAase(deoxynucleic acidase)脱氧核糖核酸酶
dNTP(four deoxynucleotide triphosphate)四种脱氧核苷酸
DOM(dissolved organic matter)溶解的有机质
D-phase(division phase)分裂期
Ds(double-stranded)双链
DSP(down stream process)下游工程
EA(elastase activity)弹性蛋白酶活性
EB(ethidium bromide)溴乙锭
EDTA Ethylene (diamine tetraacetic acid)乙二胺四乙酸
EF(elongation factor)延伸因子

EM(election microscopy)电子显微镜术
EMS(electron microscope)电子显微镜
ER(endoplasmic reticulum)内质网
ES(enzyme substrate)酶作用底物
FAD [flavin adenine dinucleotide (oxidized)]黄素酰嘌呤二核苷酸(氧化型)
$FADH_2$[flavin adenine dinucleotide(reduced)]黄素酰嘌呤二核苷酸(还原型)
FMN(flavin mononucleotides)黄素酰嘌呤单核苷酸
G(guanine)鸟嘌呤
G.R. (guaranteed reagents)保证试剂,优级纯试剂
G/Nr(glucose-nitrogen ratio)糖氮比率
GN broth(Gram-negative enrichment broth)革兰氏阴性菌增菌肉汤
GOD(glucose oxidase)葡萄糖氧化酶
G-phase[gap phase(bacterial cell cycle)]间期(细菌细胞周期)
GRAS(generally recognized as safe)一般认为是无毒的,公认无毒
GTP(guanosine-5′-triphosphate)鸟苷-5′-三磷酸
GTT(glucose tolerance test)葡萄糖耐量试验
HA(hemagglutination)血细胞凝集
Hfr(high frequency recombination)高频重组
HML(human milk lysozyme)人乳溶菌酶
HSV(herpes simplex virus)单纯疱疹病毒
I(inosine)肌苷
IEM(ion-exchange membrane)离子交换膜
Ig(immunoglobulin)免疫球蛋白
IHF(integration host factor)整合宿主因子
Inc group[incompatible group(of plasmids)]不相容群(质粒的一种特性)
IOD(immediate oxygen demand)直接需氧量,即时耗氧量
IS(insertion sequence)插入序列
ITLC(instant thin-layer chromatography)瞬时薄层色谱法
kb(kilobase)千碱基对
kDa(Kilodalton)千道尔顿
KDO(2-keto-3-deoxyoctonate)2-酮-3-脱氧辛糖酸[盐]
KDPE(2-kero-3-deoxy-6-phosphogluconate)2-酮-3-脱氧-6-磷酸葡萄糖酸[盐]
LAB(lactic acid bacteria)乳酸菌
LBP(luciferin-binding protein)荧光素结合蛋白
LC ①(lethal concentration)致死浓度;②(liquid chromatography)液相色谱法
LD (lethal dose)致死剂量
L.F.D. (least fatal dose)最小致死量
LLC(liquid-liquid chromatography)液-液色谱法
m.o.i(multiplicity of infection)感染负数

MAC(membrane-attack complex)膜侵染复合体
MCP(methyl-accepting chemotaxis protein)甲基趋化受体蛋白
MEM(minimal essential medium)基本培养基
MGF(macrophage growth factor)巨噬细胞生长因子
MHC(major histocompatibility complex)主要组织相容性复合体
MIC(minimum inhibitory concentration)最低抑制浓度
MIF(macrophage inhibition factor)巨噬细胞抑制因子
MLD ①(median lethal dose)平均致死剂量;②(minimum lethal dose)最小致死剂量
MLT(median lethal time)平均致死时间
MPD (maximum permissible dose)最大允许剂量
mRNA(messenger RNA)信使核糖核酸
MTL(mean tolerance limit)平均耐药量
MTOC(microtuble organizing centre)微管组织中心
NAD^+ (nicotinamide adenine dinucleotide)烟酰胺腺嘌呤二核苷酸,辅酶Ⅰ
$NADP^+$ (nicotinamide adenine dinucleotide phosphate)烟酰胺腺嘌呤二核苷酸磷酸,辅酶Ⅱ
NAG(N-acetyl glucosamine)乙酰氨基葡萄糖
NAM(N-acetyl muramic acid)N-乙酰胞壁酸
NB(nutrient broth)营养肉汤
NTP(ribonucleoside triphosphate)核糖核苷三磷酸
OD(optical density)光密度
OmP(outer membrane protein)外膜蛋白
ORF(open reading frame)开放阅读框
PCR(polymerase chain reaction)聚合酶链反应
PDGF(platelet-derived growth factor)血小板[源]生长因子
PEG(polyethylene glycol)聚乙二醇
PEP(phosphoenol pyruvate)磷酸烯醇式丙酮酸
Pfu(plaque-forming unit)噬斑形成单位
pg(pictogram)皮克(10^{-12} g)
PGF(platelet growth factor)血小板生长因子
PGM(phosphoglucomutase)葡糖磷酸变位酶
PHB(poly-β-hydroxybutyrate)聚 β-羟丁酸[盐]
PMF(proton motive force)质子动势
PMN(polymorphonucleoyte)多形核白细胞
PPP(pentose phosphate pathway)戊糖磷酸
Pro;P(proline)脯氨酸
PRPP(phosphoribosylpyrophosphate)磷酸核糖焦磷酸
PSⅠ和Ⅱ(photosystems Ⅰ and Ⅱ)光合系统Ⅰ和Ⅱ
PS(photosystems)光(合)系统

P-site[peptidyl site (ribosome)]肽酰基(P)位点(核糖体)

PVDF(millipore,eschborn)聚偏二氟乙烯

RBC(red blood cell)红细胞

RER(rough endoplasmic reticulum)糙面内质网

RIA(radioimmunoassay)放射免疫测定[法]

RNA(ribonucleic acid)核糖核酸

rRNA(ribosomal RNA)核糖体 RNA

RRS(Residual reduced sugar)残余还原糖

rubisco(ribulose bisphosphate carboxylase)核酮糖二磷酸羧化酶

S(svedberg coefficient)沉降系数

SDS(sodium dodecyl sulfate)十二烷基磺酸钠

SDS-PAGE(SDS polyacrylamide)SDS-聚丙烯酰胺

SEM(scanning electron microscope)扫描电镜

SHBG(sex hormone binding globulin)性激素结合球蛋白

SN(stereospecific numbering)SR 立体编号

snRNA(small nuclear ribonucleic acid)核内小 RNA

SPB(spindle pole bodies)纺锤极体

SPCA(serum prothrombin conversion accelerate)[凝血酶原] 转变加速因子,凝血因子Ⅶ

SRF,SRH(somatotropin releasing factor; somatotropin)生长激素释放因子,生长激素释放激素

SRIF(somatotropin release inhibitory factor)生长激素[释放]抑制因子

Ss(single-stranded)单链

Taq(thermus aquaticus)水生栖热菌

TCID(tissue culture infective dose)组织培养感染量

TD(tolerance dose)(可)耐量,耐药量

TEMED(N,N,N,N′-tetramethylethylene- diamine)四甲基乙二胺

TN(total nitrogen)总氮

TOC(total organic carbon)总有机碳

TOD(total oxygen demand)总需氧量

Tris(trishydroxymethylamino methane)三羟甲基氨基甲烷

tRNA(transfer RNA)转移 RNA

Trp(tryptophan)色氨酸

TSB(tryptone soya broth)大豆胰蛋白胨肉汤

U①(Unit)活力单位;②(uracil)尿嘧啶

UDP(uridine diphosphate)尿苷二磷酸

UDPG(uridine diphosphate glucose)尿苷二磷酸葡萄糖

U_L(unique long)单一长区(一串单核苷酸长链)

UOD (ultimate oxygen demand)最终需氧量,极限需氧量

U_S(unique short)单一短区(一串单核苷酸短链)

第3课 食品化学类

°T(absolute temperature)①绝对温度,开氏温度;②特尔纳度[牛奶酸度的表示法]

2,4-D(2,4-dichlorophenoxyacetic acid)2,4-二氯苯氧乙酸

5-FDUR(5-fluorodeoxyuridine)5-氟脱氧尿苷

5-FU(5-fluorouracil)5-氟尿嘧啶

5-HT(5-hydroxytryptamine;serotonin)5-羟色胺

6APA(6-aminopenicillanic acid)6-氨基青霉烷酸

A site(aminoacyl site)氨酰基部位

A. R.(analytical reagent)分析试剂,分析纯

A;Ado(adenosine)腺[嘌呤核]苷

AA(amino acid)氨基酸

AcG(accelerator globulin)促凝血球蛋白

AcOEt(ethyl acetate)乙酸乙酯

AcOH(acetic acid)乙酸

ACP ①(acid phosphatase acyl carrier protein)酸性磷酸酶酯酰载体蛋白;②(acyl carrier protein)脂酰[基]载体蛋白

ACTH(adrenocorticotropic hormone)促肾上腺皮质激素

ADP(adenosine diphosphate)腺苷二磷酸

ADPG(adenosine diphosphate glucose)腺苷二磷酸葡萄糖

AFD-G (alkali flame detector-gas chromatography)碱性火焰检测器气体色谱联合法

AFGP (antifreeze glycoprotein)抗冻糖蛋白

AFP (antifreeze protein)抗冻蛋白

AIDS(acquired immune deficiency syndrome)艾滋病

Ala;A(alanine)丙氨酸

Alp(alkaline phosphatase)碱活性磷酸酶

AM(amylase)淀粉酶

AMP(adenosine monophosphate)腺苷一磷酸

AMV(avian myeloblastosis virus)鸟类成髓细胞性白血病病毒

An. (anisol)茴香醚

ANP①(atrial natriuretic peptide)心房肽,心钠肽;②(atrial natriuretic protein)心房蛋白,心钠蛋白

AOM(active oxygen method)活性氧法

APF(animal protein factor)动物蛋白因子

APS(adenosine-5′-phosphosulfate)腺苷酰硫酸,腺苷-5′-磷酸硫酸酐

Arc(Arcylamide)丙烯酰胺

Arg;R(arginine)精氨酸

arom.(aromatic)芳香的,芳香族的

Asn;N(asparagines)天冬酰胺

Asp;D(aspartic acid)天冬氨酸

AV(acid value)酸值

BAL(dimercaprol;dimercaptopropanol)二巯基丙醇

BaP(benz[a]pyrene)苯并[a]芘

BET(brunauer-emmet-teller)单分子层

BHA(butylated hydroxyanisole)丁基羟基茴香醚

BHT(butylated hydroxytoluene)二丁基羟基甲苯

BMR(basal metabolic rate)基础代谢率

bp(boiling point)沸点

BPTI(aprotinin)牛胰蛋白酶抑制剂

BSP[brom(o)sulf(ophth)alien]四溴酚酞磺酸钠

BZOH(benzoic acid)苯甲酸,安息香酸

C.P.(chemically pure)化学纯

Cyd;C(cytidine)胞[嘧啶核]苷

cAMP(cyclic adenosine monophosphate)环腺苷酸

CANP(calcium-activated neutral proteinase)钙活性化中性蛋白酶

CAP ①(cyclic AMP receptor protein) cAMP cAMP 受体蛋白;②(catabolite gene activation protein)分解[代谢]物基因活化蛋白

CAT(chloramphenicol acetyltransferase)氯霉素乙酰基转移酶

cat.(catalyst)催化剂

CBG(corticosteroid-binding globulin)皮质脂类结合球蛋白,皮质类固醇结合球蛋白

Cbz(carbobenzoxy)苄氧羰基

CCC(chlorocholine chloride)氯化氯胆碱,矮壮素

ccc-DNA(covalently closed circular DNA)共价闭环 DNA

CCK(cholecystokinin)缩胆囊肽

CD(circular dichroism)圆二色性

CDGF(cartilage-derived growth factor)软骨[源]生成因子

cDNA(complementary DNA)互补[于 RNA 的]DNA,反转录 DNA

CDP(cytidine diphosphate)胞苷二磷酸

CEA(carcino-embryonic antigen)癌胚抗原

CEC(cation exchange capacity)阳离子交换能力

CETP;LTP-1(cholesteryl ester transfer protein)胆固醇脂转移蛋白

CF(citrovorum factor)嗜橙菌因子,亚叶酸,N^5-甲酰-5,6,7,8-四氢叶酸

CGF(chondrocyte growth factor)软骨细胞生长因子

CGN(chymotrypsinogen)胰凝乳蛋白酶原

CHR(chromatographicallly pure)光谱纯

CIM(chemical ionization mass spectrometry)化学电力质谱测定法
CLIP(corticotropin-like intermediate peptide)促肾上腺皮质激素样中间肽
CMC(carboxymethyl cellulose)羧甲[基]纤维素
CMP(cytidine monophosphate;cytidylic acid)胞苷一磷酸
CoⅠ(coenzymeⅠ)辅酶Ⅰ
CoⅡ(coenzyme Ⅱ)辅酶Ⅱ
CoA(coenzyme A)辅酶 A
coef.(coefficient)系数
Co-IF(co-initiated factor)辅起始因子
ConA(concanavalin)伴刀豆球蛋白
CPK(creatine phosphate kinase)磷酸肌酸激酶
cpm(counts per minute)每分钟计数(同位素脉冲)
CRH(corticotrophin releasing hormone)促[肾上腺]皮质[激]素释放[激]素
cRNA①(chromosmal RNA)染色体 RNA;②(complementary RNA)互补 RNA
CSF(colony stimulating factor)菌落刺激因子,集落刺激因子
CTP(cytidine triphosphate)胞苷三磷酸
CVF(cobra venom factor)眼镜蛇毒因子
Cys;C(cysteine)半胱氨酸
Cyt(cytosine)胞嘧啶
d(density)密度
D-(dextrorotatory)右旋的
D(diopter)屈光度
D.E.①(degree of esterification)酯化度;②(dextrose equivalent)糖化率
dA;dAdo(deoxyadenosine)脱氧腺苷
DAP(diaminopimelic acid)二氨基庚二酸
DBC(N^6-2-O-dibutyryl adenosine-3′,5′-monophosphate)双丁酰环腺苷酸
dC;dCyd(deoxycytidine;cytosine deoxyriboside)脱氧胞苷
DCC;DCCI(dicyclohexylcarbodiimide)二环己基碳二亚胺
dCMP(deoxycytidylic acid)脱氧胞苷酸
DCMU(3-(3,4-dichlorophenyl)-1,1-dimethylurea)二氯苯[基]二甲脲
DCU(dicyclohexylurea)二环己脲
DDD(2,2-bis(p-chlorophenyl),-1,1-dichloroethane)滴滴滴,双对氯苯基二氯乙烷
DDT(2,2-bis(p-chlorophenyl),-1,1-tichloroethane)滴滴涕,双对氯苯基三氯乙烷
DDVP(2,2-dichlorovinyldimethyl phosphate or dichlovos)2-二氯乙烯二甲基磷酸酯,敌敌畏
DFP(diisopropyl fluorophosphates)二异丙基氟磷酸
dG;dGuo(deoxyguanosine)脱氧鸟苷
dGMP(deoxyguanylic acid)脱氧鸟苷酸
DHF (dihydrofolic acid)二氢叶酸

DHFR(dehydrofolate reductase)二氢叶酸还原酶

dIMP(deoxyinosine-5′-monophosphate)脱氧次黄苷酸,脱氧肌苷酸

dITP(deoxyinosine triphosphate)脱氧次黄苷三磷酸

d-,*l*- (dextro- ,levo-)右旋 ,左旋

DL-(*dl*-)(racemic)外消旋的

DNA(deoxyribonucleic acid)脱氧核糖核酸

DNase(deoxyribonuclease)脱氧核糖核酸酶

DNP(dinitrophenol)二硝基苯酚

DO(dissolved oxygen)溶解氧

DOPA(3,4-dihydroxyphenylalanine)3,4-二羟苯丙氨酸,多巴

DPN(diphosphopyridine nucleotide)二磷酸吡啶核苷酸,辅酶Ⅰ

ds DNA(double stranded DNA)双链 DNA

ds RNA(double stranded RNA)双链 RNA

DTT(dithiothreitol)二硫苏糖醇

E ①(enzyme)酶,激素;②(ester)酯;③(evaporativity)蒸发度

EAA(essential amino acid)必需氨基酸

EAc(ethyl acetate)乙酸乙酯

EC①(effective concentration)有效浓度;②(Enzyme Commission of the International Union of Biochemistry)(国际生物化学联合会)酶委员会;③(Enzyme Commission)酶学委员会

ECD(electron capture defector)电子捕获检测器

ECG;EKG(electrocardiogram)心电图

ECM (electro chromatography)电色谱法

EDTA(editic acid; ethylenediamine tetraacetic acid; ethylened initrolotetraacetic acid)乙二胺四乙酸

ED(eletrodialysis)电渗析

EFA(essential fatty acid)必需脂肪酸

EGF(epidermal growth factor)表皮生长因子,表皮细胞生长因子

EGTA [ethyleneglycol-bis(β-aminoethyl ether)-N,N′-tetraacetic acid]乙二醇双乙胺醚-N,N′-四乙酸

ELISA(enzyme-linked immunosorbent assay)酶联免疫吸附测定[法]

EMG(electromyogram)肌电图

EMP(embden-meyerhof-parnas pathway)EMP 途径

endo PG(endo polygalacturonase)内切聚半乳糖醛酸酶

EPR(electron paramagnetic resonance)电子顺磁共振

ER(endoplasmic reticulum)肌质网

exo PG(exo polygalacturonase)外切聚半乳糖醛酸酶

FA①(fatty acid)脂肪酸;②(folic acid)叶酸

Fab(antigen binding fragment)抗原结合片段

FAD(flavin adenine dinucleotide)黄素腺嘌呤二核苷酸
FIS(far infra red spectrometer)远红外线分光计
fluor.(fluorescent)荧光的
FTR(far infra red)远红外线
FUV(far ultra-violet)远紫外线
LD_{50}(median fatal dose)50%致死剂量
FF(fat free)脱脂的
FFA(free fatty acid)游离脂肪酸,非酯化脂肪酸
FGF(fibroblast growth factor)成纤维细胞生长因子
fMet(formylmethionine)甲酰甲硫氨酸
FMN(flavin mononucleotide)黄素单核苷酸
FPLC(fast protein liquid chromatography)快速蛋白液相层析
FR(riboflavin)核黄素
FRF;FRH(follicle stimulating hormone releasing factor)促滤泡素释放素
FSH(follicle-stimulating hormone)促滤泡素
G;Guo(guanosine)鸟[嘌呤核]苷
G6PD(glucose-6-phosphate dehydrogenase)葡糖-6-磷酸脱氢酶
GABA(γ-aminobutryic Acid)γ-氨基丁酸
gal.①(galactose)半乳糖;②(gallon)加仑
GAR(glycinamide ribonucleotide)甘氨酰胺核苷酸
GC(gas capillary column)气相色谱
GDH(glutamic dehydrogenase)谷氨酸脱氢酶
GDP①(guanosine diphosphate)鸟苷二磷酸;②(guanosine-5′-diphosphate)鸟苷-5′-二磷酸
Gel(gel electrophoresis)凝胶电泳
Gel.①(gelatine)明胶,凝胶;②(gelatinous)凝胶的,胶质的
gelat.(gelatinous)凝胶的,胶质的
GFC(gel filtration chromatography)凝胶过滤色谱法
GGF(glial cell growth factor)胶质细胞生长因子
GH(growth hormone)生长激素
GHRF(growth hormone releasing factor)生长激素释放因子
GIP(gastric inhibitory polypeptide)肠抑胃肽
GLC(gas-liquid chromatography)气-液色谱法
GLC-MC(gas-liquid chromatographymass spectrometry)色谱-质谱联合法
Gln;Q(glutamine)谷氨酰胺
Glu;E(glutamic acid)谷氨酸
Gly;G(glycine)甘氨酸
GML(growth-modulating serum tripeptide)生长调节血清三肽
GMP(guanylic acid)鸟氨酸

GOT(glutamate-oxaloacetate transaminase)谷草转氨酶

GP (glycoprotein)糖蛋白

GPT(glutamate-pyruvate transaminase)谷丙转氨酶

GRH(growth hormone releasing hormone)生长激素释放激素

GRIF(SRIF)(growth hormone release inhibitory factor)生长激素释放抑制因子

GRIH(growth hormone release inhibitory hormone)生长激素释放抑制激素

GSH(glutathione)谷胱甘肽;还原型谷胱甘肽

GSSG(glutathione)氧化型谷胱甘肽

Gua(guanine)鸟嘌呤

HAA(hepatitis associated antigen)肝炎相关抗原,澳大利亚抗原

Hb(hemoglobin)血红蛋白

HCG(human chorionic gonadotrophin)人绒毛膜促性腺激素

HDL(high density lipoprotein)高密度脂蛋白

HEPES(N′-2-hydroxyethylpiperazine-N′-2-ethanesulfonic acid)N-2-羟乙[基]哌嗪-N′-2-乙烷酸

HF(Hageman factor)哈格曼因子,凝血因子Ⅻ,接触因子

HFCS(high-fructose corn syrup)高果糖浆

HFGS(high-fructose glucose syrup)高果葡萄糖浆

Hfr(high frequency recombination)高频重组

HGF(hyperglycemic factor)高血糖因子

HGPRT(hypoxanthine guanine phosphoribosyl-transferase)次黄嘌呤鸟嘌呤转磷酸核糖[基]酶

His;H(histidine)组氨酸

HLB(hydrophile-lipophile balance)亲脂-亲水平衡

HMC(hydroxy methyl cellulose)羧甲基纤维素

HMG(human menopausal gonadotrophin)人绝经期促性腺激素

HMG-CoA(β-hydroxy-β-methylglutaryl-CoA)β-羟[基]-β-甲基戊二酸单酰辅酶 A

HMM(hexamethylolmelamine)六羟甲[基]三聚氰胺

HMP(hexose monophosphate pathway)磷酸己糖途径,磷酸己糖支路

HnRNA(heterogeneous nuclear RNA)不均一核 RNA

HPLC (high performance liquid chromatography)高效液相色谱法

Hyp(hydroxyproline)羟脯氨酸

IAA(indole-3-acetic acid)吲哚-3-乙酸

IBP(iron-binding protein)铁结合蛋白

ICSH(LH)(interstitial cell-stimulating hormone)间质细胞刺激素,黄体生成激素

IDP(inosine-5′-diphosphate)次黄嘌呤-5′-二磷酸

LDL(low-density lipoprotein)低密度脂蛋白

I.E.P. (iso-electric point)等电点,等电离点

LLE(liquid-liquid extraction)液-液萃取

IEC(ion-exchange chromatography)离子交换色谱法
IF(initiation factor)起始因子
IGF;NSILA(insulin-like growth factor)胰岛素样生长因子,不被压制胰岛素样活性多肽
IGLC(inverse gas-liquid chromatography)逆气-液色谱法
Ile;I(isoleucine)异亮氨酸
IMP①(inosine-5′-monophosphate)次黄嘌-5′-磷酸;②(inosinic acid)次黄苷酸,肌苷酸
IP_3(inositol triphosphate)三磷酸肌醇
IPTG(isoprophl-*β*-*D*-thiogalactoside)异丙基-*β*-*D*
IR(infra-red)红外线的
IRS(infra-red spectroscopy)红外线光谱学
IRSP(infra-red spectrometer)红外线分光计
ISP(infra red spectrophotometer)红外线分光光度计
ITP(inosine triphosphate)次黄苷三磷酸
IU (international unit)国际单位
JH(juvenile hormone)保幼激素
K(equilibrium constant)平衡常数
kcal(kilocalorie)千卡(热量单位)
K_{cat}(catalytic constant)催化常数,转换数
K_m(Michaelis constant)米氏常数
K_s(substrate constant)底物常数
L/S ratio(ratio of linoleic acid to saturated acid)脂肪中亚油酸对饱和脂肪酸之比
LAAO(*L*-amino acid oxidase)*L*-氨基酸氧化酶
Lac(lactose)乳糖
Leu(leucine)亮氨酸
LHFGS(high fructose glucose syrup)高果糖浆
LP(lipoprotein)脂蛋白
LPS(lipopolysaccharide)脂多糖
LS(lactose synthetase)乳糖合成酶
LSM(low sodium milk)低钠牛奶
LTH(luteotropic hormone)催乳激素
LY(lactoalbumin-yeastolate)乳白蛋白-酵母酶
Lys;K(lysine)赖氨酸
M(molecular weight)分子量
Mb(myoglobin)肌红蛋白
MC(methyl cellulose)甲基纤维素
MD(malic dehydrogenase)苹果酸脱氢酶
MDR(minimum daily requirement)最低的日需要量
Met;M(methionine)甲硫氨酸
MHb(methemoglobin)正铁血红蛋白

MS(mass spectrometer)质谱仪
MSG (monosodium glutamate)谷氨酸钠
MSNF(milk solids-not-fat)非脂乳固体
N(theoretical plate number)理论级数[色体色谱分析]
NAA①(neutron activation analysis)中子活化分析;②(nicotinic acid amide)烟酰胺
NBOD(nitrogenous biochemical oxygen demand)含氮生化需氧量
NDEA(nitrosodiethylamine)二乙基亚硝胺
NDMA(nitrosodimethylamine)二甲基亚硝胺
NMR(nuclear magnetic resonance)核磁共振
non-vol. (non-volatile)非挥发性
NPip(nitrosopiperidine)亚硝基哌啶
NPN(non-protein nitrogen)非蛋白氮
NPU(net protein utilization)净蛋白质利用率
NR(nutritive ratio)营养价值
O.D.①(optical density)光密度;②(outside diameter)外径;③(oxygen demand)需氧量
O/W (oil in water)水包油
O/W emulsion(oil in water emulsion)水包油乳浊液
OAR(organic analytical reagent)有机分析试剂
OAS(organic analysis standard)有机分析标准
OD unit(optical density unit)光密度单位
OHV(hydroxyl value)羟值
OI(oxygen index)氧指数
O-R(oxidation-reduction)氧化-还原
OR(optical rotation)①旋光性;②旋光度
ORD(optical rotatory dispersion)旋光色散
org. (organic)有机的
ORP(oxidation-reduction potential)氧化-还原电位
ORT(oxidation-reduction titration)氧化-还原滴定(法)
Osm(osmol)渗透压单位
P. (poise)泊[黏度单位]
P.E. (protein equivalent)蛋白质当量
P.N. (peroxide number)过氧化值
PABA(*p*-aminobenzoic acid)对氨基苯甲酸
PAL (pyridoxal)吡哆醛,维生素 B_6醛
PAM(pyridoxamine)吡哆胺,维生素 B_6胺
PAS(para-aminosalicylic acid)对氨基水杨酸
PC(polycarbonate)聚碳酸酯
PC(PPC)(paper chromatography)纸上色层析法
PCBs(polychlorinated biphenyls)多氯联苯

PCE(paper chromatoelectrophoresis)纸色层电泳法
PE (pectin esterase)果胶酯酶
PER(protein efficiency ratio)蛋白质效率
PGA(pteroylglutamic acid)叶酸
PGL(polymethylgalacturonatelyase)聚甲基半乳糖醛酸裂解酶
PHA[phytoh(a)emagglutinin]植物凝集素
Phe(phenylalanine)苯基丙氨酸
PIN(pyridoxine)吡哆醇,维生素 B_6
PMG(polymethylgalacturonase)聚甲基半乳糖醛酸酶
poly- (polymer)聚合物
polym.①(polymerize)聚合;②(polymorph)多晶型物
POV(peroxide value)过氧化物价
PP(inorganic pyrophosphate)无机焦磷酸盐
Ppc(paper partition chromatography)纸分区色谱法
PPi(inorganic pyrophosphate)无机焦磷酸
ppn(precipitation)沉淀(作用)
PQI(protein quality index)蛋白质质量指数
prim. (primary)初级,最初的,主要的
Pro(proline)脯氨酸
PSP(phenolsulfonphth alein)酚红
PUFAs(polyunsaturated fatty acids)多不饱和脂肪酸
PyP(pyrophosphate)焦磷酸酯
Q_{10}(temperature coefficient)温度系数
RDNA(recommended daily nutrient allowance)推荐每日营养需要量
Redox(reduction-oxidation)氧化-还原
RNAase(ribonucleic acidase)核糖核酸酶
RP-HPLC(reversal phase-high pressure liquid chromatography)反相高压液相色谱
S.C.(standard condition)标准状态[指温度压力]
S.D.S.dev.(standard deviation)标准误差,均方误差,标准偏差
S.E.N.(steam emulsion value)蒸汽乳化值
S.N.(saponification number)皂化值
S.T.(Surface Tension)表面张力
SAA(surface acting agent)表面活性剂
Sap(saponification)皂化(作用)
sap.val(saponification value)皂化值
sat.(saturate)饱和
sat.sol.(saturated solution)饱和溶液
satd.(saturated)饱和的
SCE(standard calomel electrode)标准甘汞电极

SCR(semi conservative replication)半保留复制
SDS(sodium dodecyl sulfate)十二烷基磺酸钠
SE(saponification equivalent)皂化当量
Ser(serine)丝氨酸
SNF(solid non fat)(牛奶的)非脂肪固体
SNP(soluble nucleoprotein)可溶性核蛋白
SOD(superoxide dismutase)超氧化物歧化酶
SP.V.(specific volume)比容
SRF(somatotropin releasing factor)生长激素释放因子
SS ①(solid state)固态;②(suspended solids)悬浮固体
st.s.(standard sample)标准样品
STD(standard test dose)标准试验剂量
STI(soybean trypsin inhibitor)大豆胰蛋白酶抑制剂
stzn(sterilization)杀菌,消毒
SW(salt water)盐水
T①(thymine)胸腺嘧啶;②(trace)痕量,微量
T°①吉尔尼尔度[牛奶酸度的表示法];②克氏绝对温度
T_0[ice point (absolute ℃)]冰点
$t_{1/2}$(half-life)半衰期
TAN(total ammonia nitrogen)总氨态氮
Tc(critical temperature)临界温度
TCA①(trichloroacetic acid)三氯乙酸;②(tricarboxylic acid)三羧酸
TGase(trans glutaminase)转谷氨酰胺酶
TH(thyrotropic hormone)促甲状腺激素
THFA(tetrahydrofolic acid)四氢叶酸
ThG(thyroglobulin)甲状腺球蛋白
Thr(threonine)苏氨酸
Thy(thymine)胸腺嘧啶
TLC(thin layer chromatography)薄层色谱法
TMP(thymidine-5′-monophosphate)胸苷-5′-磷酸
TPN(triphosphopyridine nucleotide)三磷酸吡啶核苷酸
tRNA(transferRNA)转移 RNA
Try(tryptophan)色氨酸
TSH(thyroid-stimulating hormone)促甲状腺激素
TSS(total suspended solids)悬浮总固体
TTA(total thermal analysis)总热分析
TTP(thymidine-5′-triphosphate)胸苷-5′-三磷酸
TVN(total volatile nitrogen)总挥发氮
TVP(textured vegetable protein)组织化植物蛋白

Tyr(tyrosine)酪氨酸

U(uracil)尿嘧啶

U.S.RDA(United States recommended da daily allowance)美国建议每日容许量[包括维生素、蛋白质、矿物质等]

UDP(uridine-5′-diphosphate)尿苷-5′-二磷酸

USP(up stream process)上游工程

UTP(uridine-5′-triphosphate)尿苷-5′-三磷酸

USA(ultraviolet spectral analysis)紫外线光谱分析

UV(ultraviolet light)紫外线

V-B(vitamin B)维生素 B

V-C(vitamin C)维生素 C

Xan;X(xanthine)黄嘌呤

XMP(xanthylic acid)黄苷酸

αFP;AFP[α-f(o)etoprotein]甲胎蛋白

第 4 课　食品包装类

ACC(activated calcium carbonate)活性碳酸钙

AP(aseptic packaging)无菌包装

B.B.(base box)基箱[镀锡薄板的包装单位]

BI(behavior identity)活动知识

BIB(bag in box)衬袋纸箱

BIC(bag in carton)衬袋纸盒

BOPP(both oriented polypropylene)双向拉伸聚丙烯

CA(cellulose acetate)乙酸纤维素

CAP(controlled-atmosphere storage and packaging)气调包装

CD(corona discharge)电晕放电

CF(consumption factor)消耗因子

CIS(corporate identity system)企业形象识别系统

CTI(critical temperature indicators)临界温度指示卡

CTTI(critical temperature/time integrators)临界温度-时间指示卡

EAA(ethylene aorylic acid copolymer)乙烯-丙烯酸共聚物

EMA(equilibrium modified atmosphere)平衡调节气体

EP(epoxy resin)环氧树脂

EVA(ethylene-vinyl acetate copolymer)乙烯-乙酸乙烯共聚物

EVOH(ethylene-vinyl alcohol copolymer)乙烯-乙烯醇共聚物

FCM(food contact materials)食品接触材料

GARS(generally recognized as safe)公认安全级

HDPE(high density polyethylene)高密度聚乙烯

HDPE-HM(high density and high molecular polyethylene)高密度高分子聚乙烯

HDPE-MM(high density and medium molecular polyethylene)高密度中分子聚乙烯

HFFS(horizontal form-fill-seal)卧式成型-充填-封口包装

HIPS(high impact polystyrene)耐冲击聚苯乙烯

HMPE(high molecule polyethylene)高分子聚乙烯

HR(Rockwell hardness)洛氏硬度

HRP(heat resistance plastic)耐热塑料

LCA(life cycle assessment)生命周期分析和评估法

LDPE(low density polyethylene)低密度聚乙烯

LHDPE(linear high density polyethylene)线性高密度聚乙烯

LLDPE(linear low density polyethylene)线性低密度聚乙烯

LMDPE(linear medium density polyethylene)线性中密度聚乙烯

LSFO(least shelf-life first out)最短货架期法

MAP(modified-atmosphere storage and packaging)气调包装

MDPE(medium density polyethylene)中密度聚乙烯

MI(mind identity)理念识别

MVTR(moisture vapour transmission organisms)水蒸气迁移速率

NMBP(non-migratory bioactive polymers)非迁移生物活性聚合物

OP①(oriented polystyrene)定向聚苯乙烯;②(osmotic pressure)渗透压

OPA(ONY)(oriented polyamide)单向拉伸聚酰胺

OPP①(oxygen partial pressure)氧气分压;②(oriented polypropylene)拉伸聚丙烯;③(oriented polypropylene)单向拉伸聚丙烯

PA (polyamid)聚酰胺

PAN(polyacrylonitrile)聚丙烯腈

PC(polyoarbenate)聚碳酸酯

PCR(post-consumer recycle)回收利用

PCTFE(polychlorotri-fluoroethylene)聚三氟氯乙烯

PE (polyethylene)聚乙烯

PEG(polyethylene glycol)聚乙二醇

PEN(polyethylene naphthalate)聚乙烯石脑油

PEs(polyester)聚酯

PET(polyethylene terephthalate)聚对苯二甲酸乙二醇酯[聚酯];聚乙烯对苯二酸聚酯

PHA_S(polyhydroxy alkanoates)多羟基链烷酸酯

PIB(polyisobuylene)聚异丁烯

PP(polypropylene)聚丙烯

PPO(polyphenol oxidase)多酚氧化酶

PS(polystyrene)聚苯乙烯

PSB(polybutadiene-styrene)聚丁二烯-苯乙烯

PTFE(polytetrafluoroethylene)聚四氟乙烯

PTLF(preformed tray and liddirgfilm)托盘充填封口包装

PTR(Association of Packaging Technology and Research)包装技术协会

PUR(polyurethane)聚氨酯

PVA(polyvinyl alcohol)聚乙烯醇

PVAC(polyvinyl acetate)聚乙酸乙烯酯

PVC(polyvinyl chloride)聚氯乙烯

PVDC(polyvinylidene chloride)聚偏二氯乙烯

RFID(radio frequency identification)射频识别

SA(styrene-acrylonitrile clpolyemer)苯乙烯-丙烯腈共聚物

SGS(soluble gas stabilization)可溶气体稳定化

SLDS(shelf life decision system)货架期判断系统

SMAS(safety monitoring and assurance system)安全检测和保证系统

SML(specific migration limit)特别的迁移限量

SSO(specific spoilage organism)特定腐败菌

ST(snorkel type)充气包装

STP(see-through packaging)可见包装

TFFS(thermo-form-fill-seal)热成型-充填-封口包装

TMA(trimethylamine)三甲胺

TPS(thermoplastic starch)热塑性淀粉

TTI(time-temperature indicators)时间-温度指示卡

ULDPE(ultra low density polyethylene)特低密度聚乙烯

UNIQUAC(universal quasi-chemical activity)通用类化学活性

V.D. (vapor density)蒸汽密度

VC(vacuum chamber)真空包装

VFFS(vertical-form-fill-seal)立式成型-充填-封口包装

VFH(vacuum film handing)真空薄膜处理

VI(visual identity)视觉识别

VLDPE(very low density polyethylene)极低密度聚乙烯

第 5 课 与食品有关的组织和标准类

AAFCO(Association of American Food Control Official)美国饲料控制官员协会

ACh.S. (American Chemical Society)美国化学学会

ACRI(Air-Conditioning and Refrigeration Institute)空气调节与制冷学会

ACS ①(American Chemical Society)美国化学学会；②(American Crystal Sugar Company)美国结晶糖公司

ACSIRO(Australian Commonwealth Scientific and Industrial Research Organization)

澳大利亚联邦科学及工业研究组织

ADA(American Dairy Association)美国乳品工业协会

AFDO(Association of Food and Drug Officials)食品与药品官员联合会

AFFI(American Frozen Food Institute)美国冷冻食品研究所

AIN(American Institute of Nutrition)美国营养学会

AMS(Agricultural Marketing Service)农业市场管理局

AOAC(Association of Official Agriculture Chemists)官方农业化学家协会

APA(Administrative Procedure Act)行政程序法

APEC(Asia-Pacific Economic Cooperation)亚太经济合作组织

APHIS(Animal and Plant Health Inspection Service)动植物健康检验局

APLAC(Asia Pacific Laboratory Accreditation Cooperation)亚太实验室认可合作组织

ARS(Agricultural Research Service)农业研究管理局

AS ①(American Standard)美国标准;②(anabolic steroid)蛋白同化激素

ASB(American Society of Biochemists)美国生化学家协会

ASBC ①(American Society of Biological Chemists)美国生物化学家协会;②(American Society of Brewing Chemists)美国酿造化学家协会

ASM(American Society for Microbiology)美国微生物协会

ASRE(American Society of Refrigerating Engineers)美国制冷工程师协会

ASTM(American Society for Testing and Materials)美国试验和材料协会

AVC(Association of Vitamin Chemists)美国维生素化学工作者协会

B.A.S. (British Association Standard)英国标准协会

B.A.U. (British Association Unit)英国标准单位

B.F.M.I.R.A. (British Food Manufacturing Industries Research Association)英国食品工业研究协会

B.P. (British Pharmacopoeia)英国药典

B.V.U. (British Viscosity Unit)英国黏度单位

BCS(Bureau of Chemistry and Soils)化学与污染局

BESA(British Engineering Standards Association)英国工程标准协会

BMS(British Mycological Society)英国真菌协会

BPF(British Plastic Federation)英国塑料联盟

BS(British Standard)英国标准

BSE(bovine spongiform encephalopathy)牛海绵状脑病/疯牛病

BSI(British Standard Institute)英国标准协会

BSS(British Standard Specification)英国标准技术条件[规范]

C.F.T.A. (Canadian Food Technologists' Association)加拿大食品工艺师协会

C.F.T.R.I. (Central Food Technological Research Institute)中央食品工艺研究所

C.W.U. (Chemical Workers Union)化学化工者联合会

CAC(Codex Alimentarius Commission)食品法典委员会

CACEB(China National Accreditation Committee for EMS Certification Bodies)中国

环境管理体系认证机构认可委员会

Can. Pat(Canadian Patent)加拿大专利

CC(Codex Committee)法典委员会

CCFAC(Codex Committee on Food Additive and Contaminants)食品添加剂与污染物食品法典委员会

CCFH(Codex Committee on Food Hygiene)食品卫生法典委员会

CCPR(Codex Committee on Pesticide Residues)农药残留法典委员会

CDC(Center for Disease Control & Prevention)美国疾病预防控制中心

CEPM(Committee of Experts on Phytosanitary Measures)植物卫生措施专家委员会

CFIA(The Canadian Food Inspection Agency)加拿大食品检验局

CFR(Code of Federal Regulations)美国的联邦管理法典

CFSAN(Center for Food Safety and Applied Nutrition)食品安全与应用营养中心

CGSB(Canadian General Standards Board)加拿大通用标准局

CMA(Cigar Manufacturers Association)雪茄烟制造商协会

CNAB(National Accreditation Board of Certifiers)中国认证机构国家认可委员会

CNAL(China National Accreditation Committee for Laboratories)中国实验室国家认可委员会

CNAT(China National Auditor and Training Accreditation Board)中国认证人员与培训国家认可委员会

COAB(Canadian Organic Advisory Board)加拿大有机咨询委员会

COG(Canadian Organic Growers)加拿大有机栽培者协会

CQC(China Quality Certification Center)中国质量认证中心

CSIRO ①(Commonwealth Scientific & Industrial Research Organization)联邦科学与工业研究组织；②(Commonwealth Scientific & Industrial Research Organization, Australia)澳大利亚联邦科学与工业研究组织

CSREES(Cooperative State Research, Education, and Extension Service)加州合作科研、教育及推广管理局

CVM(Center for Veterinary Medicine)兽药中心

CVMP(Committee for Veterinary Medicinal Product)兽药委员会

Dan. p(Danish Patent)丹麦专利

DAP(Deutsches Apothekerbuch)德国药典

DDF(Danish Directorate for Fisheries)丹麦渔业委员会

DEFRA(Department for Environment Food and Rural Affairs)环境、食品与农村事务部

DFIC(Dehydrated Foods Industry Council)脱水食品工业委员会

DPD(Danish Plant Directorate)丹麦植物委员会

DSB(Dispute Settlement Body)(世贸组织)争端解决机构

DVFA(Danish Veterinary and Food Administration)丹麦兽医与食品安全管理局

EEC(European Economic Community)欧洲经济共同体

EFSA(European Food Safety Authority)欧盟食品安全管理局

EMEA(The European Agency for the Evaluation of Medicinal Product)欧洲药物评价局

EPA(Environmental Protection Agency)国家环境保护局

EPIA(Egg Products Inspection Act)蛋类产品检验法

ERS(Economic Research Service)(美国农业部)经济研究局

EU(European Commission)欧盟

F.P.(French Patent)法国专利

FACA(Federal Advisory Committee Act)联邦顾问委员会法案

FACC(Food Additives and Contaminants Committee)(英国)食品添加剂及污染物委员会

FAO[Food and Agriculture Organization(of the United Nations)]联合国粮食与农业组织

FAR(Food Additives Regulation)(美国)食品添加剂规范

FDA(Food and Drug Administration)(美国)食品药品监督管理局

FFDCA(Federal Food,Drug, and Cosmetic Act)联邦食品、药物、化妆品法

FHCCAC(The Food Hygiene Committee of the Codex Alimentation Commission)食品卫生委员会

FMF(Food Manufacturers' Federation)(英国)食品制造商联合会

FMIA(Federal Meat Inspection Act)联邦肉类检验法

FNB(Food and Nutrition Board)食品和营养委员会

FOIA(Freedom of Information Act)自由信息法案

FQPA(Food Quality Protection Act)食品质量保护

Fr.P.(French Patent)法国专利

FS(Federal Specification)(美国)联邦规格

FSA(Food Standard Agency)食品标准署法

FSAI(The Food Safety Authority of Ireland)爱尔兰食品安全局

FSANZ(Food Standards Australia/New Zealand)澳大利亚/新西兰食品标准局

FSC(Food Standard Committee)食品标准委员会

FSIS(Food Safety and Inspection Service)食品安全及检验局

FSQS(Food Safety and Quality Service)食品安全与质量管理处

FVO(Food & Veterinary Office)(欧盟委员会)食品和兽医办公室

GAO(General Accounting Office)美国国家审计署

GAP(Good Agricultural Practices)良好农业规范

GCSL(Gulf Coast Seafood Laboratory)海湾水产品实验室

Ger.Pat(German Patent)德国专利

GRAS(Substances Generally Recognized as Safe)一般认为安全的物质

GVP(Good Veterinarian Practices)良好兽医规范

IAF(International Accreditation Forum)国际认可合作组织

IAMS(International Association of Microbiological Societies)国际微生物学会联合会

ICS(Indian Chemical Society)印度化学学会

ICUMSA(International Commission for Uniform Methods of Sugar Analysis)国际食糖统一分析法委员会

IDA(International Dietetic Association)国际营养师协会

IDC(International Documentation Centre)国际文献资料中心

IDF(International Dairy Federation)国际乳品联合会

IEC(International Electrotechnical Commission)国际电工委员会

IFJU(International Fruit Juice Union)国际果汁联合会

IFOAM(International Federation of Organic Agriculture Movements)国际有机农业运动联盟

IFT(Institute of Food Technologists)食品工艺家学会

IIR(International Institute of Refrigeration)国际制冷技术学会

IIS(Integrated Inspection System)综合检查系统

ILAC(International Laboratory Accreditation Conference)国际实验室认可论坛

INACOL(Institute National I'Ame'lioration des Conserves de Le'gumes)比利时全国罐头食品工业研究所

IOCU(International Organization of Consumers Unions)国际消费者联盟组织

IOFI(International Organization of the Flavor Industry)国际食用香料工业组织

IP(International Pharmacopoeia)国际药典

IPPC(International Plant Protection Convention)国际植物保护公约

IPRO(International Patent Research Office)国际专利研究会

IRRI(International Rice Research Institute)国际稻米研究所

ISO ①(International Standard Organization)国际标准组织；②(International Organization for Standardization)(联合国)国际标准化组织

ITRI(International Tin Research Institute)国际锡研究所

IUB(International Union of Biochemistry)国际生物化学联合会

IUC(International Union of Chemistry)国际化学联合会

IUF(International Union of Food Workers)国际食品工作者联合会

IUFoST(International Union of Food Science and Technology)国际食品科学技术联合会

IUNS(International Union of Nutritional Science)国际营养科学联合会

Jap. P. (Japanese Patent)日本专利

JAS(Japanese Agricultural Standard)日本农林规范

JECFA(Joint FAO/WHO Expert Committee on Food Additives FAO/WHO)食品添加剂联合专家委员会

JECFI(Joint Expert Committee on Food Irradiation)食品辐射联合专家委员会

JIS(Japanese Industrial Standard)日本工业标准

ILSI(international Life Science Institute)国际生命科学研究所

JMPR(Joint FAD/WHO Meeting on Pesticide Residues)农药残留专家委员会联席会议

JP(Japanese Pharmacopoeia)日本药典

JPI(Japan Packaging Institute)日本包装研究所

JSFA(Japanese Standard of Food Additives)日本食品添加剂标准

ISO(International Standards Organization)国际标准化组织

ISR(International Standards Research)国际标准研究所

M.O.H.(Ministry of Health)卫生部

MAFF(Ministry of Agriculture ,Fisheries and Food)(英国)农业,渔业和食品部

MOF(Ministry of Food)食品工业部

MRL(Maximum Residue Limits)最高残留限量

N.A.S.(National Academy of Sciences)(美国)国家科学院

N.C.A.(National Canners Association)(美国)国家罐头协会

N.Z.P.(New Zealand Patent)新西兰专利

NACA(National Agriculture Chemical Association)(美国)全国农业化学协会

NACMCF(The National Advisory Committee on Microbiology Criteria for Food) 美国食品微生物标准咨询委员会

NAFTA(North American Free Trade Agreement)北美自由贸易协议

NASA(National Aeronautics and Space Administration)美国宇航局

NBS(National Bureau of Standard)(美国)国家标准局

ND ①(National Distillers and chemical Corporation)(美国)国家酿酒化学公司;②(not detected)未检出

NEN(Netherlands Norm)荷兰国家标准

NF(Norm Francaises)法国标准

NFA(National Food Authority)国家食品局

NIH(National Institutes of Health)国家健康研究所

NMFS(National Marine Fisheries Service)国家海洋渔业局

NOAA(National Oceanic and Atmospheric Administration)国家海洋大气管理局

NP ①(Norwegian Patent)挪威专利;②(Norma Portugal)葡萄牙国家标准;③(nucleoprotein)核蛋白

NS ①(normal saline)标准盐水;②(Norsk Standard)挪威标准

NSA(National Standards Association)(美国)国家标准协会

NSDA(National Soft Drink Association)(美国)全国清凉饮料协会

NSPA(National Soybean Processors' Association)(美国)全国大豆加工商协会

OECD(Organization for Economic Cooperation and Development)国际经济合作与发展组织

OFDC(Organic Food Development and Certification Center)中国国际有机食品认证中心

OHSMS(Occupational Safety and Health Management of System)职业健康与安全管理体系

OIE(Office International des Epizooties)国际动物卫生组织/国际兽疫局
PAC(Pacific Accreditation Cooperation)太平洋认可合作组织
PCR(polymerase chain reaction)聚合酶链反应
PHSA(Public Health Service Act)公共健康事业法案/公共卫生服务法
PI(Pharmacopoeia International)国际药典
PIBAC(Permanent International Bureau of Analysis Chemistry)国际分析化学常设局
PIC(Prior Informed Consent Procedure)事先知情程序
PL(positive list)(食品添加剂的)准许使用名单
PN(Polska Norma)波兰国家标准
PPIA(Poultry Products Inspection Act)禽类产品检验法
QA(quality assurance)质量保证
QC(quality controlling)质量控制
SAC(Society for Analytical Chemistry)分析化学协会
SAFPHC(Safety Assessment on Food Processing and Hazard Control)食品加工的安全评估和危害控制
SPHE(Society of Packaging ,Handling and Engineer)(美国)包装与装卸工程师协会
SPS(Sanitary and Phytosanitary Standard)动植物卫生检疫措施
SSOP(Sanitation Standard Operating Procedure)标准卫生操作规程
TBT(Technical Barriers to Trade)贸易技术壁垒协定
TCFA(Technical Consultation on Food Allergies)食物过敏技术咨询委员会
TIRC(Tobacco Industry Research Committee)(美国)烟草工业研究委员会
TQM(total quality management)全面质量管理
U.S. Pat(United States Patent)美国专利
UNDP(United Nations Development Programe)联合国开发计划署
UNEP(United Nations Environment Programmer)联合国环境署
UNFA(United Nations Food and Agriculture Organization)联合国粮食及农业组织
UNIDO(United Nations Industrial Development Organization)联合国工业发展组织
USAID(United States Agency for International Development)美国国际开发署
USDA(United States Department of Agriculture)美国农业部
USP(United States Pharmacopoeia)美国药典
WHO(Word Health Organization)世界卫生组织
WTO(Word Trade Organization)世界贸易组织

附录

作业参考答案

第1章

1. Effect of retrograded resistant starch types on forming rice starch gel

Abstract: In order to improve the structure and qualities of general rice starch products, effects of retrograded resistant starches (RSⅢ) made from *Castanea henryi* (*C. henryi*), potato and mung bean on microstructure and physicochemical properties of rice starch gels were investigated by using modern analysis methods, including RVA-4 rapid viscosity analyzer, S-3 400N scanning electron microscope, TA-XT Plus structure analyzer, D/max 2 500 automatic X-ray diffractometer and DSC200 differential scanning calorimeter. The results demonstrated that the structure and properties of rice starch gel had a significant change ($P<0.01$) after adding *C. henryi*, potato and mung bean RSⅢ, respectively. Especially the role of *C. henryi* RSⅢ was the most prominent. The *C. henryi*, potato and mung bean RSⅢ had no effect ($P>0.05$) on viscosity properties of rice starch pastes. Rice starch gel without RSⅢ had a lot of irregular, deep and shallow large holes, while the net structure of rice starch gel with RSⅢ became more uniform and dense. RSⅢ had little influence on gumminess and cohesiveness of rice starch gel ($P>0.05$). The *C. henryi*, potato and mung bean RSⅢ could accelerate the formation of rice starch gel. Compared with the rice starch gel without RSⅢ, their hardness increased by 2.38, 1.97 and 1.25 times ($P<0.01$), stickness increased by 2.56, 1.99 and 1.32 times ($P<0.01$), and spring increased by 1.07, 0.81 and 0.53 times ($P<0.01$), respectively. A-type crystal was dominant in rice starch, and V-type crystal was dominant in *C. henryi* RSⅢ, while B-type crystal was dominant in potato and mung bean RSⅢ. Whether or not adding RSⅢ, V-type crystal in rice starch gel powders was transformed as the main crystal shape, and their total relative crystallinities had not changed ($P>0.05$). In addition to a low temperature endothermic peak, rice starch pastes with RSⅢ showed a high temperature endothermic peak. Whether or not to add RSⅢ, the change of temperature parameters of low temperature endothermic peak was a little ($P>0.05$), but its endothermic enthalpy decreased significantly ($P<0.01$). As for the high temperature endothermic peak, all parameters of rice starch pastes with potato and mung bean RSⅢ had no difference ($P>0.05$), but they were significantly higher than those of rice starch paste with *C. henryi* RSⅢ ($P<0.01$). To sum up, adding RSⅢ with different sources can effectively improve the structure and quality of rice starch gel. The present results can provide an important reference for using the RS to improve the qualities and nutritional functions of the rice products.

2. Effect of controlled freezing point preservation on structure and functional characteristics of beef myofibrillin

Abstract: The influence of conrtolled freezing piont storage(−1 ℃) on the structural and functional properties of beef myofibrillar protein (MP) were investigated by determining ATPase activity, sulfhydryl groups, thermal stability and SDSPAGE analysis, and samples stored at 4 ℃ were used as the control. The results showed that the functional properties of MP were significantly decreased with increasing storage time and SDS-PAGE model showed that MP was also degraded. In the same storage time, however, the changes of functional properties could be reduced by superchilling beef compared with chilling beef. For 12 day storage, the Ca^{2+}-ATPase activities of chilling and superchilling beef decreased by 79.2%and 46.9%. Total sulfhydryl contents of chilling and superchilling beef decreased by 50. 12% and 35. 08%, reactive sulfhydryl contents of chilling and superchilling beef decreased by 93.68% and 71.15%. The protein solubility of chilling and superchilling beef MP decreased by 41. 18% and 15. 52%, respectively. SDS-PAGE analysis revealed that MP showed more severe degradation at 4 ℃ than that at −1 ℃. Meanwhile, the thermal stability of MP stored at −1 ℃ was higher than that at −1 ℃. Therefore, superchilling is an effective methods to preserve freshness and quality of beef.

3. Effect of polymerized whey protein concentrate on the stability of fermented milk beverage

Abstract: The objective of this study was to determine the effect of different amounts of polymerized whey protein concentrate on the stability of fermented milk beverage. The effects of the addition of polymerized whey protein concentrate on the centrifugal sedimentation rate, light absorption ratio, particle size, viscosity, surface hydrophobicity, free thiol groups of fermented milk beverage were investigated. The results showed that the centrifugal sedimentation rate of fermented milk beverage with polymerized whey protein concentrate was significantly lower than that of fermented milk beverage without polymerized whey protein concentrate($P<0.05$). With the increase of the polymerized whey protein concentrate, light absorbing of fermented milk drink down slightly after the first increase, viscosity of fermented milk drinkincreased gradually, the particle size decreased slightly and then increased slightly, and the addition amount of the polymerized whey protein concentrate was 30% when the size of the system was smallest, the surface hydrophobicity and free thiol groups ofsystem increased gradually, and the free thiol groups of the fermented milk drink with 10%-40% polymerized whey protein concentrate was significantly higher than that of 0%($P<0.05$). These results indicated that adding appropriate polymerized whey protein concentrate is conducive to the stability of fermented milk beverage.

4. Effect of dietary supplementation of tartary buckwheat on physiological metabolism and intestinal flora in mice with high-fat diet induced dyslipidemia

Abstract: The objective of this work was to ascertain the effect of dietary supplementation of tartary buckwheat on the physiological metabolism and intestinal flora of dyslipidemic mice. For this purpose, a mouse model of dyslipidemia was created by feeding mice a high-fat diet (HFD). Changes in serum lipid metabolism-related indicators and the intestinal flora were determined respectively by high performance liquid chromatography and the plate count method and were correlated with each other. The results showed that HFD induced lipid metabolic disturbance and that both protein and starch from tartary buckwheat significantly reduced serum lipid levels in dyslipidemic mice to almost normal values ($P < 0.05$). The number of beneficial intestinal bacteria (Bifidobacterium, Lactobacillus, and Enterococcus) in the tartary buckwheat starch and protein intervention groups was significantly higher than in the HFD group ($P < 0.05$), and the number of harmful bacteria (Escherichia coli) was significantly lower ($P < 0.05$). Furthermore, correlation analysis indicated that both tartary buckwheat starch and protein could adsorb bile acid and cholesterol and consequently promote their excretion, thereby regulating serum lipid metabolism and the intestinal microbial flora balance. Meanwhile, the increased proportion of intestinal priobiotic bacteria and the decreased proportion of harmful bacteria promoted lipid metabolism modulation and oxidative stress inhibition. Therefore, tartary buckwheat regulates blood lipid metabolism most likely by promoting bile acid salt excretion, regulating the intestinal flora balance, and improving oxidative stress.

第 2 章

1. The distinguishing factors that affect the efficiency of modification are the starch source, amylose to amylopectin ratio, granule morphology, and type and concentration of the modifying reagent. The extent of alteration in the starch properties reflects the resistance or the susceptibility of starch towards different chemical modifications. Modified starches with desirable properties and degree of substitution can be prepared by critically selecting a suitable modifying agent and a native starch source.

2. In the fats and oils industry, deacidification of oils is important not only for consumer acceptance but also because it has the maximum economic impact on production. Chemical, physical, and miscella deacidification methods have been used in the industry. There are several drawbacks associated with these conventional deacidification processes. Some new approaches that may be tried out are biological deacidification, reesterification, solvent extraction, supercritical fluid extraction, and membrane technology.

3. Iron is a mineral that is necessary for producing red blood cells and for redox processes. Iron deficiency is considered to be the commonest worldwide nutritional deficiency. The most effective technological approaches to combat iron deficiency in developing countries include supplementation targeted to high risk groups combined with a program of food fortification and dietary strategies designed to maximize the bio-availability of both the added and the intrinsic food iron.

4. High pressure (HP) treatment has emerged as a food processing technology primarily due to increasing interest in the preservation of foods. Applying HP to food products modifies interactions between individual components, influences rates of enzymatic

reactions, and can inactivate microorganisms. HP induced changes in rennet coagulation time, rate of curd formation, and cheese yield in cheese process.

5. It is recognized that chemicals from packaging and other food-contact materials can migrate into the food itself and thus be ingested by the consumer. The monitoring of this migration has become an integral part of ensuring food safety. This article reviews the current knowledge on the food safety hazards associated with packaging materials together with the methodologies used in the assessment of consumer exposure to these hazards.

6. This review discusses the current approach and status of the methodologies used to evaluate the effect of high pressure pasteurization and sterilization on food safety and quality aspects. It was investigated whether the existing approaches for process impact evaluation of conventional thermal pasteurization and sterilization are transferable to the assessment of high pressure high temperature process impact and what the points of particular attention are.

7. The increasing interest of consumers in functional foods has brought about a rise in demand for functional ingredients obtained using "natural" processes. In this review, new environmentally clean technologies for producing natural food ingredients are discussed. This work provides an updated overview of the principal applications of two clean processes, supercritical fluid extraction, and subcritical water extraction, used to isolate natural products from different raw materials, such as plants, food by-products, algae, and microalgae.

8. The industry of fresh-cut fruits and vegetables is constantly growing due to consumers demand. New techniques for maintaining quality and inhibiting undesired microbial growth are demanded in all the steps of the production and distribution chain. The use of ultraviolet-C, modified-atmospheres, heat shocks, and ozone treatments, alone or in different combinations have proved useful in controlling microbial growth and maintaining quality during storage of fresh-cut produce.

9. The Maillard reaction begins with a condensation reaction between the carbonyl group of a reducing sugar with the amine group from an amino acid or protein, developing into a complex group of reactions. The Maillard reaction can be divided into three stages: early, advanced, and final stages. A final phase leads to the formation of nitrogenous and/or brown-colour polymers, known as melanoidins. The Maillard reaction occurs in food processing and storage, and has some effects on the physical, chemical, biological and organoleptic characteristics of food products in which it occurs.

第3章

1. In this paper, the effects of blanching, dehydration method, and temperature on certain characteristics of okra were investigated. Fresh okra fruits contained a higher level of pigments and viscosity than dehydrated products. Compared with unblanched samples, samples blanching prior to dehydration retained more of the color components but were

less viscous. Samples dehydrated under vacuum condition retained more ascorbic acid, pigment, and mucilage at each of the dehydration temperatures than those from a hot air oven. High dehydration temperatures have a negative effect on the color, ascorbic acid, and viscosity of samples.

2. An attempt has been made to concentrate anthocyanin, a natural red colorant from red radish using an integrated membrane process, which involves ultrafiltration(UF), reverse osmosis(RO) and osmotic membrane distillation(OMD). A comparison of UF, RO, and OMD was made when they are operated individually and in combination with each other. It was observed that, the integrated membrane processes have the advantages of achieving a higher concentration of anthocyanin compared to that of the individual membrane processes, which resulted in an increase in the concentration of anthocyanin from 40 mg/100 mL to 980 mg/100 mL.

3. The ability of biodegradable films to prolong the shelf life of minimally process lettuce stored at 4 ℃ was addressed. Four different films were tested: two polyester-based biodegradable films(NVT1, NVT2), a multilayer film made by laminating an aluminum foil with a polyethylene film (All-PE), and an oriented polypropylene film(OPP). Package headspace, microbial load, and color of the packed lettuce were monitored for a period of 9 days. Results show that the gas permeability of the investigated films plays a major role in determining the quality of the packed produce. Moreover, it was observed that biodegradable films guarantee a shelf life longer than that of OPP.

4. In this work, sterilization of a viscous liquid food in a metal can lie horizontally and rotated axially in a still retort was simulated. The rotating speed set at 10 r/min and samples were heated by steam at 121 ℃. Transient temperature and velocity profiles caused by natural and forced convection heating were presented and compared with those for a stationary can. The results indicated that the combined effect of natural and forced convection splits the slowest heating zone(SHZ) into two distinct regions, unlike what has been previously observed in the stationary can. The volume of the SHZ was found to cover less than 5% of the total volume of the rotated can at the end of heating, which is due to the effect of rotation.

5. In this study, the freezing time and rate of 1 cm^3 cauliflower floret samples were determined under different freezing conditions in an air blast freezer. Four different air temperatures(−20, −25, −30 and −35 ℃) and six different air velocities(70, 131, 189, 244, 280 and 293 m/min) were applied in the freezer, and the freezing rate and time of cauliflower pieces were determined under each condition. The freezing time of cauliflower samples frozen with cold air at −20 ℃ and 280 m/min was similar to that of samples frozen with cold air at −35 ℃ and 70 m/min. When the velocity of air was increased from 70 m/min to 293 m/min, the freezing time was approximately halved.

6. Most existing additives and all new ones must undergo an extensive toxicological evaluation to be approved for use. Although questions continue to be asked regarding the

validity of animal studies, there is a consensus among scientists that animal testing does provide the information needed to make safety decisions.

7. Inside the cells of fruits and some vegetables, there are phenolic compounds that are colorless in the fresh fruit, but which turn brown or greyish when oxidized. These phenolic substances are the substrate inside the cells on which the oxidative enzymes are present. The discoloration results when the enzyme catalyzes the oxidation of these phenolic substances inside the cells when in the presence of air. It can be prevented by inactivating the enzymes or by excluding oxygen. Antioxidants are also effective in preventing enzymatic browning because they are more readily oxidized than the substrates (phenolic compounds) hence the antioxidants they can be used to protect the oxidation of the phenolic compounds.

8. A nutritious diet has five characteristics. One is adequacy: the foods provide enough of each essential nutrient, fiber, and energy. Another is balance: the choices do not overemphasize one nutrient or food type at the expense of another. The third is calorie control: the foods provide the amount of energy you need to maintain appropriate weight—not more, not less. The fourth is moderation: the foods do not provide excess fat, salt, sugar, or other unwanted constituents. The fifth is variety: the foods chosen differ from one day to the next.

第 4 章

1. Strawberry fruits were treated for 30 min in 50 kV/m or 100 kV/m high-voltage electrostatic field(HVEF), and then stored for 12 days at 4-8 ℃ and RH 90%. The results showed that the decay index, MDA content, PPO activity, anthocyanin content of strawberry fruits decreased, treatment of 100 kV/m HVEF being better. However, HVEF was not good to keep the appearance quality of the fruits.

2. Soybean peptides with DH of about 24% were produced using Alcalase and Neutrase and were fractionated into 4 fractions using ultrafiltration membranes with 30 000, 10 000 and 5 000 of molecular weight cutoff, respectively. Such properties of the 4 fractions were examined as solubility, foaming property and foam stability, emulsifying property and emulsifying stability and antioxidative activity, and ACE inhibitory activity.

3. With the rapid development of GMOs, more and more GMO foods have been pouring into the market. Much attention has been paid to GMO labeling under the controversy of GMO safety. Transgenic corns and their parents were scanned by a continuous wave of near-infrared diffuse reflectance spectroscopy range of 12 000-4 000 cm^{-1}; the resolution was 4 cm^{-1}; scanning was carried out for 64 times; Near-infrared diffuse reflectance spectroscopy is unpolluted and inexpensive, so it is a very promising detection method for GMO foods.

4. The determination methods of the freeze-tolerant ability of bakeries yeast were discussed. By comparison, a remarkable positive correlation was found among yeast cell

survival ratio after freezing, the relative values of fermentation ability after freezing in liquid fermentation(LF) medium and the dough were analyzed critically. The evaluation using the LF medium was easy to operate and control, and could be applied to yeast screening and tests on a large scale.

5. The main components in hazed apple juice concentrate were analyzed and compared with the normal juice concentrate. The contents of total phenolic compounds, condensed tannins, soluble protein, and metal irons such as K^{+}, Ca^{2+}, Mg^{2+} in the hazed juice concentrate were higher than those in the normal juice concentrate, while the content of small molecular phenolic compounds was lower.

6. The change of physicochemical properties of potato starch was studied after microwave radiation to potato starch with a moisture content of 30%. By microwave treatment, a concavity in the surface could be observed clearly. The swelling power, solubility, and syneresis decreased. The X-ray intensities of the major d-spacing increased and the X-ray pattern of potato starch changed from B type to A type. The pasting temperature rose, the viscosity dropped and the viscosity curve pattern changed from A type to C type. Besides, starch molecules decomposed to some extent.

7. Solid phase microextraction(SPME) is a new extraction technique based on solid phase extraction(SPE). The system consists of sampling, extraction and condensation in a unit and has the advantage of simple operation, rapid analysis, low cost, safety, good resolution, and high sensitivity. The apparatus, principle, and operation of the SPME system were summarized. The optimization of its working parameters and conditions, and its applications in food analysis were reviewed. Its future development was also proposed.

8. *Streptococcus thermophilus* and *Lactobacillus bulgaricus* are commonly used in yogurt fermentation. They are very difficult to separate due to symbiosis. A comparison was taken between plate count and test-tube count using modified Elliker medium and acidified MRS medium. A new test-tube count was adopted to count *Lactobacillus bulgaricus* during yogurt fermentation. The results showed that *Lactobacillus bulgaricus* and *Streptococcus thermophilus* could be separated effectively by the new test-tube count. The test-tube count is more convenient and has lower contamination risk during inoculation and cultivation. In addition it also provides higher accuracy, repeatability, and detection rate than the plate count. Therefore the test-tube count is an easy and effective way to count *Lactobacillus* spp. It may be also applied in isolation and detection for other facultative anaerobes.

第5章

1. The native mixed linkage β-glucan of cereals is classified as a soluble dietary fiber, with physiological properties generally similar to guar gum and other random coil polysaccharides. The ability of oat and barley products to attenuate postprandial glycemic and insulinemic response is related to the content of(1,3)(1,4)-β-*D*-glucan(β-glucan) and

viscosity. A role of the viscosity of β-glucan has not been directly demonstrated for lowering of serum cholesterol levels, and not all studies report a statistically significant lowering.

2. Maltodextrin, a partially hydrolyzed starch, is composed of a mixture of amylose and amylopectin, both having a different influence on hygroscopicity and tablet properties such as tensile strength, friability and disintegration time. Therefore, this report describes the influence of the maltodextrin grade(different amylose/amylopectin ratios) on powder hygroscopicity, flowability, density, and compactibility.

3. Continuous-flow microwave pasteurization systems have created much interest in the beverage industry and thus have been investigated for the various beverages, such as apple juice, milk, and orange juice. While many investigations on microwave pasteurization have focused on microbial and enzyme inactivation, very little work has been conducted on evaluating process and product parameters of a continuous-flow microwave pasteurization system. Therefore, the objectives of this project were to design a lab-scale continuous flow microwave pasteurization system for apple cider and to characterize the processing parameters.

4. Foodborne illnesses are an overwhelming public health problem in the US. Education and training of food handlers are critical since worker mishandling could cause the outbreaks of foodborne illnesses. Currently, hundreds of food safety educational materials are available; Unfortunately, differences in cultural, economic and social factors associated with workers themselves make it difficult to use the same educational food safety program in all situations. Additionally, educational materials may not be effective if they are designed without looking at the worksite social, physical, and environmental factors surrounding the target audience.

第6章

1. A novel method of DNA extraction and purification was developed and was used in conjunction with a multiplex real-time PCR assay for the simultaneous detection of *Salmonella* and *Listeria monocytogenes* in a raw meat sample. The PCR used primers targeting the *inv*A gene of *Salmonella* and the *hly*A gene of *L. monocytogenes*, and PCR products were detected with a Light Cycler on the basis of fluorescence from SYBR Green and melting temperature. The assay allowed the detection of 2 *Listeria* cells and 4 *Salmonella* cells per g of the original sausage within 10 h, including an enrichment period of 6 to 8 h.

2. The effect of processing on functional compounds in buckwheat was investigated. Extractions of buckwheat flour were carried out before and after roasting or extrusion. Folin-Ciocalteu assays indicated that processing did not cause any change in total phenolic content in buckwheat flour. Roasted(200 ℃, 10 min) dark buckwheat flour exhibited an increase in non-polar compounds as well as polar compounds whereas extrusion exhibited

increase only in polar compounds. Antioxidant activity test(DPPH) showed that roasting at 200 ℃ for 10 min decreased the antioxidant activity slightly whereas extrusion(170 ℃) did not cause any change. The results suggest that processing conditions can be optimized to retain the health promoting compounds in buckwheat products.

3. Nonstarch polysaccharides(NSPs), both soluble and insoluble, were added to pasta doughs at levels of 2.5%, 5%, 7.5%, and 10% levels. The cooking and textural characteristics of the pasta were evaluated using a range of analytical techniques. Generally, NSP addition was found to increase the cooking losses, and reduce the protein and starch contents of the pasta. This effect was dependent on the level of NSP added and also the type(soluble or insoluble). Pasta firmness was generally reduced in relation to the level of NSP addition, although some gel-forming NSPs resulted in higher firmness values. Pasta stickiness, adhesiveness, and elasticity were also affected. The results indicate that careful selection of NSP addition is needed to ensure optimum textural and cooking characteristics in NSP enriched pasta products.

第7章

1. This study investigated the impact of kilning on α-amylase, β-amylase(total and soluble), β-glucanase, and protease activities in buckwheat malt. Common buckwheat (*Fagopyrum esculentum*) was steeped at 10 ℃ for 12 h, germinated at 15 ℃ for 4 days, and kilned at 40 ℃ for 48 h. Moisture content and enzymatic activities were determined throughout the kilning period. Results showed moisture content was reduced from 44% to 5% after 48 h of kilning at 40 ℃.

2. Compositional changes of fruit pulp and peel during ripening of white-and pink-fleshed guava fruits were studied. The white and pink guava fruits exhibited a typical climacteric pattern of respiration. Fruit tissue firmness decreased progressively, in a similar manner, in both guava fruit types. Total soluble solids(TSS) and total sugars increased in pulp and peel of both guava types with a decrease in flesh firmness.

3. The antioxidant activity of pomegranate juices were evaluated by four different methods(ABTS, DPPH, DMPD, and FRAP) and compared to those of red wine and a green tea infusion. Commercial pomegranate juices showed an antioxidant activity(18-20 TEAC) three times higher than those of red wine and green tea(6-8 TEAC). The activity was higher in commercial juices extracted from whole pomegranates than in experimental juices obtained from the arils only(12-14 TEAC).

4. Conventionally, thermal processing almost inactivated all the microorganisms and pectolytic enzymes and produced microbial and consistently stable tomato juices; however, they also reduced the color, extractable carotenoids and lycopene and vitamin C compared with fresh juice. During storage, all the pressure processing could improve the extractable carotenoids and lycopene contents compared with fresh juice, and they also retained more vitamin C contents than thermal processing.

5. Patulin is an important mycotoxin in apples and apple products and it is also a marker of quality in the apple and apple juice industry. Numerous methods are described in the literature regarding its extraction and analysis from clear and cloudy apple juice as well as liquidized solid apples. However, there is a dearth of information concerning patulin analysis in dry solid apple products e. g. apple rings which cannot be liquidized. We developed a method to solve this problem and validated it for precision, accuracy, and linearity at 10, 30, and 50 ppb. The method is based on solid phase extraction and isocratic separation on HPLC-DAD.

第8章

1. Whey protein concentrates are commonly used as ingredients in numerous foods because of their excellent technological functionality and high nutritional value. They are applied in liquid food preparations as textural ingredients to increase firmness or cause gel formation after heating. Besides, several sources have suggested specific physiological properties of whey protein. Undenatured whey protein is now being investigated as a dietary aid for the enhancement of immune status through intracellular glutathione synthesis.

2. The presence of ethylene in the storage atmosphere has been shown to decrease the storage life of several fruit and vegetables, as well as increasing decay and the incidence of some physiological disorders. In the case of climacteric fruit, exogenous ethylene accelerates ripening, which is usually undesirable in stored fruit. It is suggested that the concentration of ethylene in the storage environment can be directly related to the rate of quality loss in a wide range of fruit and vegetables.

3. Food-borne outbreaks resulting from consumption of *Listeria*-contaminated foods continue to raise a major concern with regard to food safety. In the United States, *Listeria monocytogenes* was responsible for about 25% of the estimated food-borne-disease-related death. Milk and milk products are frequently incriminated. Consequently, considerable effort has been made to control the pathogen in dairy products. Among dairy products, yogurt received the least attention due to the fact that its high acidity and milk pasteurization were thought to be effective barriers to the growth of many pathogens including *L. monocytogenes*.

4. Native from the Americas, sweet bell pepper is a Solanaceous fruit, whose consumption is growing in popularity, mainly due to its occurrence in a wide variety of colours(ranging from green, yellow, orange, red, and purple), shapes, and sizes and its characteristic flavour. Bell peppers are used to produce dehydrated products (such as paprika), pickled peppers, and sliced or diced frozen peppers to be used in pizzas or to be eaten raw as salads. The demand for sliced and diced frozen raw peppers has been increasing considerably in the last years, due to consumers' willing to eat raw, minimally processed vegetable products, as part of healthier food habits.

5. Biodegradable films are generally made from biopolymers such as carbohydrates, proteins, and lipids. Films made from carbohydrates have good mechanical properties and they are also effective barriers against low polarity compounds, but these coatings do not offer a good barrier against humidity. On the other hand, protein-based films are generally excellent barriers to oxygen, carbon dioxide, and some aromatic compounds, but their mechanical properties are not satisfactory, which limits their possible use in different applications.

6. Cheese is a fermented milk product. The basic principle of making cheese is to coagulate the milk so that it forms into curds and whey. Today's methods help the curdling process by the addition of a starter(a bacterial culture) and rennet the coagulating enzyme which speeds the separation of liquids(whey) and solids(curds). Generally, cheese making starts with acidification. This is the lowering of the pH of the milk. Classically, this process is performed by bacteria, which feed on the lactose in milk and produce lactic acid as a waste product.

7. Nowadays, many food companies offer almost ready to eat processed potato products(i. e., fry strips, hash browns or chips) which are consumed widely across the world. Such products are ready to eat prior to a short time of frying, microwave heating, cooking, or directly after pouring of boiling water. Plant tissue subjected to these simple culinary processing undergoes physicochemical changes that are decisive for the texture formation and texture properties of the end-use product. The texture(i. e., structure, thickness, firmness, crispiness, color, taste, etc.) is of critical importance to the consumer's choice and acceptability as well as the direct factor affecting the product's quality.

8. Recent studies on the total antioxidant capacity of fruits and vegetables revealed that a large group of colourful compounds, some of which are flavonoids (including anthocyanins, flavones, isoflavones, catechin, and epicatechin), may be responsible for antioxidant protection against free radicals. In addition to the vitamins and minerals known to be present in fruits and vegetables, phytochemicals such as flavonoids and other phenolic may decrease the risk of cardiovascular disease and various forms of cancer by providing enhanced antioxidant protection in the human body.

9. Milk can be fermented to a variety of different products varying in appearance, flavor, texture, and health beneficial properties. The art of producing such different products is based on different technologies and in the selection of the lactic acid bacteria used in the fermentation process. The emergence of new technologies such as molecular biology and genetics advanced on the one hand broaden the possibilities to use these techniques for selection to screen thousands of natural strains and mutants, and on the other hand, to improve strains by direct genetic engineering. This article illustrates examples of both applications in starter strain development for dairy fermentations. Furthermore, it partitions the nature of genetic modifications into three groups: (ⅰ) one-

step genetic events like deletions, gene amplification, plasmid insertions or losses; (ⅱ) multi-step genetic rearrangements with DNA of the same species; and (ⅲ) trans-species genetic modifications. The nature of the genetic alteration has an impact on the safety assessment and legal implications in different countries.

10. The Center for Food Safety is a national non-profit membership organization committed to protecting human health and the environment by promoting organic agriculture and other sustainable practices. CFS engages in legal initiatives, grassroots mobilizations, and education programs designed to influence government and industry and to inform the public on such issues as genetic engineering, food irradiation, and organic food standards.

11. Wine is a complicated chemical liquid, the nature of which scientists are only just beginning to understand. It is estimated that wine contains more than 1 000 volatile flavor compounds, of which more than 400 are produced by yeasts. Despite decades of research, scientists are only now beginning to get a fuller understanding of the true nature of the chemical composition of wine. Part of the difficulty is that the picture is a dynamic one. The volatility of various flavor compounds can be altered by other components of the wine.

12. The foods with the highest contribution to acrylamide intake vary from country to country depending on national dietary patterns and methods of food preparation in a particular country. In general, potato products, coffee, and bakery products are the most important sources. In Germany, for example, bread and bread rolls account for 25% of the acrylamide intake due to the high consumption of almost 240 g/day.

第9章

1. Riemann H, Cliver D. Foodborne infections and intoxications. ISBN: 978-0-12-588365-8.2005.

2. Nutrient-Dense Emergency Relief Food Product. Washington, D. C: Institute of Medicine National Academy Press.

3. Zhao Siming, Yu Lanling, Xiong Shanbai, et al. Study on Rheological Properties of Rice Amylopectin. The Chinese society of agricultural machinery,2003(3):58-60.

4. Whitehurst J R. Emulsifiers in Food Technology.

5. Abraham A M. Sustainability Science and Engineering.

6. Defining Principles. ISBN: 978-0-444-51712-8. 2006.

7. Experts from Dole Food Company. Encyclopedia of Foods. A Guide to Healthy Nutrition. ISBN: 978-0-12-219803-8.2002.

8. Brostoff J, Challacombe S. Food allergy and intolerance. 2002. ISBN:978-0-7020-2038-4.

9. An international multi-disciplinary journal for the dissemination of food-related research consumer food safety: British Food Journal, 0007-070X; V.107, No. 7,2005.1

10. Luo Denglin, Nie Ying, Zhong Xianfeng. Ultrasound-assisted extraction of

ginsenosides from ginseng in supercritical CO_2, The Chinese Society of Agricultural Engineering, 2007(6):256-258.

11. Xu Keyong, Wu Cai'e, Li Yuanrui, et al. Supercritical carbon dioxide extraction of garlic oil from garlic juice. The Chinese Society of Agricultural Engineering, 2005, 21(4):150-154.

12. Luo Denglin, Qiu Taiqui, Wang Ruirui. Effect of ultrasound on ginsenosides extraction by supercritical CO_2 reverse microemulsions. Journal of JangShu University (Natural Science Edition), 2006, 27(3):202-206.

13. Yuan Yiquan. Theory and Application of Ultrasound in Modern. Nanjing: Nanjing University Press, 1996.

14. Liao Chuanhua, Huang Zhenren. The technology of supercritical fluid extraction, craft development, and its application. Beijing: Chemistry Industry Press. 2004:28-29.

15. Hu Fei, Chen Ling, Li Lin, et al. Study on rheological properties of micronized potato starch. The Chinese Cereals and Oils Association, 2003, 18(2):61-63.

16. Hu Fei, Chen Lin, Li Lin. Konjac variation of crystal structure of potato starch in the process of micronization milling. Fine Chemicals, 2002, 19(2):114-117.